John Collins Warren

The Healing of Arteries After Ligature in Man and Animals

John Collins Warren

The Healing of Arteries After Ligature in Man and Animals

ISBN/EAN: 9783337241476

Printed in Europe, USA, Canada, Australia, Japan

Cover: Foto ©berggeist007 / pixelio.de

More available books at **www.hansebooks.com**

THE

HEALING OF ARTERIES

AFTER LIGATURE

IN MAN AND ANIMALS

BY

J. COLLINS WARREN, M. D.,

ASSISTANT PROFESSOR OF SURGERY, HARVARD UNIVERSITY. SURGEON TO THE MASSACHUSETTS
GENERAL HOSPITAL. MEMBER AMERICAN SURGICAL ASSOCIATION. HONORARY
FELLOW PHILADELPHIA ACADEMY OF SURGERY.

NEW YORK
WILLIAM WOOD & COMPANY
1886

The Publishers

Book Composition and Electrotyping Co.

157 and 159 William St., New York.

Was die Erfindung der Buchdruckerkunst für die Wissenschaft, was die Erfindung des Schiesspulvers für den Krieg, was die Erfindung der Eisenbahn für den Verkehr der Völker untereinander, das ist die Erfindung der Arterienunterbindung für die Chirurgie.

DIFFENBACH, Die Operative Chirurgie,
Leipzig, 1845, i., 121.

PREFACE.

THE study of the subject of which this monograph treats, has been carried on, hitherto, in a more or less fragmentary way, different portions of it having received very minute attention, and from the hands of the ablest pathologists. The attempt has been made here to study the question from a more comprehensive standpoint, to observe not only the behavior of the various tissues concerned in the process of repair, but also the different phases through which the vessel passes from the moment of ligature until the condition is reached after which no further change occurs.

The investigations have, for the most part, been carried on in the Harvard Medical School, and the writer takes this opportunity to express his appreciation of the great facilities offered for such work in the Physiological and Pathological Laboratories.

Experiments were performed also in the Veterinary Department of the University, with the permission of Professor C. P. Lyman.

Through the courtesy of Dr. J. S. Billings, the very valuable collection of arteries in the Army Medical Museum, at Washington, was placed at the writer's disposal for study.

Finally the writer wishes to express his indebtedness to Dr. H. P. Quincy for skilled assistance in many of his experiments.

BOSTON, *March*, 1886.

CONTENTS.

THE LIGATURE OF ARTERIES.

CHAPTER I.

HISTORY.

LISFRANC well says: "La ligature a été * * * l'objet de recherches si multipliées que l'histoire des diverses procédés des expériences tentées pour connaître la manière d'agir des instruments inventés pour la faciliter ou l'améliorer pourrait fournir matière à plusieurs volumes." It needs but a casual glance at the literature of this important department of surgery to discover that the ligature was employed by surgeons in early historic times.

Probably the first recorded use of a ligature is that mentioned by Súsrutas' 1500 B.C., who employed it in tying the umbilical cord: no mention, however, is made by him of its use in surgery, nor do we find any in the writings of the early Egyptians, from whom the Greek writers of that time obtained their knowledge of surgery. Had the ligature been in use at that time we should have undoubted evidence thereof in the writings of Hippocrates [2] (460-375 B.C.). By some writers he has indeed been regarded as the discoverer of the ligature. The passage upon which this claim is based is the following: "Sanguinem e venis profluentum sistunt animi deliquium, figura aliorsum tendens venæ interceptio, linamentum contortum appositio deligatio."

This rendering of the Greek text is not accepted by some of the best authorities, and the sixth book, in which it occurs, is regarded by many as not genuine.*

Praxagoras (335 B.C.) first discovered the difference between

---

* Both Haeser [26] and Greifenberger state emphatically that no traces of the mention of the ligature are to be found in the writings of the Hippocratic school. In Adams' "Genuine Works of Hippocrates," there is no allusion to it.

I

arteries and veins, and his pupil, Herophius, and a colleague, Era-
sistratus, first bound limbs previous to amputation, to prevent
hemorrhage.

It is probable that the ligature was first introduced by some surgeon of the Alexandrian school, in which great progress was made
in the study of anatomy and surgery; for Celsus,[3] (25-30 B.C. 45-50
A.D.) who was deeply versed in the literature of his time, and particularly that of Alexandria, derived his knowledge of the ligature
from this School.

Aulus (or Aurelius Cornelius) Celsus was one of the first Roman
medical authors. The following quotation from his works gives a
fair idea of the means used to arrest hemorrhage at that day.

"If the hemorrhage be alarming, which may be known by the
situation and size of the wound, and from the violence of the bleeding, the wound is to be filled with dry lint, and a sponge squeezed
from cold water is to be pressed firmly on it by the hand. If the
bleeding does not subside by these means, the pledgets are to be frequently changed, and if not sufficiently powerful whilst dry, they
are to be moistened with vinegar. This last is a powerful agent for
suppressing hemorrhage; and, on that account, some pour it into
the wound; but here again it is to be feared that the matter, by
being powerfully retained there, may subsequently produce high inflammation. It is this which prevents one using corrosives, or those
applications which, by their caustic quality, induce an eschar; although most of them check hemorrhage. However, if once in a
way we do have recourse to such, the mildest are preferable. But
should these means fail also, the bleeding vessels should be taken
up, and, ligatures having been applied above and below the wounded
part, the vessels are to be divided in the interspace, that thus they
may retract while their orifices remain closed. When the case does
not admit of this measure, the vessels should be cauterized."

It is supposed that the following passage from the writings of
Celsus indicates that linen thread was the material used; "Qua parte
vero inhærebunt et ab superiore et ab inferiore parte lino vinciendæ,
etc." The manner of applying the ligature is thus described: "But,
before excision, these at their extremities ought to be tied with a
thread, its ends being left out of the wound, like as in other veins
requiring ligature."

In amputations, the ligature does not appear to have been used,
probably because Celsus was not familiar with the anatomy of
arteries. When these vessels, after being severed, had retracted

beneath the surface of the wound, he did not attempt to find them and tie them.

The successors of Celsus say little about the ligature, a circumstance supposed to be due to the fact that they did not think it necessary to mention this detail of an operation.* It is doubtful if the ligature were much used. Amputation was avoided as much as possible.

Archigenes,[5] however, who flourished about the end of the first century of the Christian Era, made such modifications in the methods of amputation that he was able to resort to it much more freely than other surgeons of his time. He controlled bleeding by bandages placed around the limb, applied provisional ligatures to large vessels, and tied others after the limb had been removed.

Rufus of Ephesus recognized the difference between arteries and veins. Both he and Heliodorus, a colleague of Archigenes, were familiar with torsion, as is shown in the following quotation from Rufus: "Vas immissa volsella extendemus et moderate circumflectemus, at ubi ne sic quidem cessaverit (hemorrhagia) vinculo constringemus."

Heliodorus[6] says in a description of an operation for the radical cure of hernia, "After laying bare the tunica dartos, the larger vessels should be tied; the smaller ones should be transfixed with a hook twisted round several times, and by means of the twisting closed."

Claudius Galenus[7] (131-211 A.D.) makes mention of the ligature in many parts of his works. He was, however, not a practical surgeon, and avoided the use of the knife. According to some authorities his imperfect knowledge of the vessels led him to recommend tying the central end of the vessel only. According to Adams he advises an accurate examination of every deep-seated vessel to ascertain whether it is an artery or a vein, "after which it is to be seized with a hook, and twisted moderately. If the flow of blood is not stopped thereby, he recommends us, if the vessel is a vein, to endeavor to restrain it without a ligature by the use of styptics, or things of an obstruent nature, such as roasted rosin, the fine down of wheaten flour, gypsum, and the like. But, if the vessel is an

---

* Haeser quotes a passage from Paulus Ægineta in support of this view which he renders thus: "Having first secured, of course, (wie natürlich, ὡς εἰκός) the vessels." The same passage is, however, rendered by Adams: "Securing, as is proper, with a thread, any vessel that may come in the way."

artery, he says one of two things must be done, either a ligature must be applied to it, or it must be cut across. He adds, we are even obliged sometimes to apply a ligature to large veins and cut them across." He used silk thread and thread of Celtic linen, and also gut, not in the Lister sense, but simply for the purpose of securing a durable material; he even mentions the place where he obtained his ligatures, "The shop on the Via Sacra between the temple Roma and the Forum." In this connection the fact is interesting that in the Naples Museum are to be seen a pair of sliding forceps found at Pompeii, which were evidently intended to be used with the ligature. Galen was afraid that an aneurism might develop if the ligature did not come away before the wound was closed by new growth. He thought that the vessels were healed by a growth from the surrounding tissues; "quæ namque caro in abscisis vasorum partibus coalescit, ea pro operculo est ac osculum eorum claudit."

Antyllus, at the end of the third century, not only invented an operation for the cure of aneurism which, with slight modification, still bears his name; but he also speaks of the same method of dealing with vessels in removing tumors. He says: "If the vessel cannot be pushed aside, it should be tied on both sides of the wound and cut through; if the tumor were under the carotid or jugular it would be unsafe to try this method, as it might produce instant death."

Among the latest of the Greek writers are Aetius[8] (502-575), and Paulus Ægineta[9] (625-690); the latter mentions the ligature more frequently than any other ancient writer. "When the bleeding is stopped," he says, "endeavor, if it is a vein, to restrain the blood without a ligature by the same medicine; but if it is an artery one of two things must be done,—either apply a ligature around it or cut the vessel asunder, by which means you will restrain the blood. Sometimes too we are obliged to apply a ligature to large veins, and also occasionally to cut them asunder transversely. * * * You may know whether it is a vein or an artery that pours forth the blood, from this: that the blood of an artery is brighter and thinner, and is evacuated by pulsation, whereas that of the vein is blacker and without pulsation. The following passage has special significance, showing that the ligature was not confined to small vessels before the time of Paré. "But if the weapon has lodged in any of the larger vessels, such as the internal jugular or carotid and the large arteries in the armpits or groins, and if the extraction threaten a great hemorrhage, they are first to be secured with ligatures on both sides, and then the extraction is to be made." This practice was, however, limited to the class of injuries mentioned, and rarely extended to amputation and other surgical operations. Torsion is not mentioned by Paulus, although it was in use in the time of Oribasus (326-403) who, in a quotation from Heliodorus, describing the radical cure of hernia, states that after laying bare the tunica dartos, the larger vessels are to be tied, the small ones to be seized with a hook, and twisted round several times, and by this means closed. It is probable that, in later times, the more frequent use of the actual cautery displaced this method.

The Arabian physicians, who transmitted the writings of these authors to the Italian and French surgeons, although they mentioned the ligature, rarely used it, confining themselves almost entirely to the actual cautery. During the period at which this school flourished, little or nothing was done for the advancement of surgery. This was due to the insufferable prejudices of that time: religion forbade the study of anatomy, and the medical man regarded it as a disgrace to be employed in any manual labor, and contented himself with writing prescriptions, leaving the use of the cautery and the knife to subordinates. Avenzohar [13] (1113-1162) says: "Omnia hace ad dictos servitores medicorum habent pertinere ad medicum autem honoratum nihil aliud pertinere dicimus nisi ut consilium præstet solummodo ciborum medicinarumque infirmi absque aliqua operatione manuum quemadmodum non convenit ei facere sirupos et electuaria suis manibus."

Avicenna [14] (980-1037) also says of amputations of the arm and thigh, "et medicus debet ab eo fugere." Rhazes (850-922) used linen thread ligatures, and recommends them for large vessels, but he generally advises the more fashionable styptic and cautery. Avicenna has but little to say about the ligature, which he confines to arteries alone.

Albucasis or Abulcasis [14] (1106), the most prominent man of this school, was more familiar with the writings of Paulus Ægineta, and used the ligature more freely, although he devoted a whole book to the cautery. He thus describes the ligature: "At sin arteria magna sit oportet illam in duobis locis ligare cum filo duplicato forti, sit vero filum ex serico vel ex testudinis chordis ne festinet illis corruptio antequam vulnus consolidetur, accideret enim hemorrhagia. Tum amputa, quod superfluum est, inter duas ligaturas." Before removing tumors he tied the large vessels, excising the growth a day or two later. He performed the operation of Antyllus for aneurism.

Neither Avenzohar, nor his pupil, Averroes,[15] the later representatives of the Arabian school, has much to say about the ligature.

The Italian school naturally suffered from the influence of its predecessor: medical literature was studied only by the clergy, and the religious prejudices of that time presented the same obstacles to the advance of surgical science that it had to the Arabian school. The chief improvement was the introduction of the mediate ligature applied with a needle, like a stitch, by Roger of Parma (1214), as recorded by his pupil Roland; and his example was followed by almost all other Italian surgeons. Bruno (1252), gives the advice, when other means fail, to seize the artery with a hook, and lift it up before passing the ligature. In Lanfranchi's Surgery,[16] which was written in the middle of the tenth century, occurs the following passage: "Oportet te tunc aut venam ligare et ipsam de loco extrahere et caput venæ vel arteriæ contorquere aut ferro candente sanguinem sistere," showing that the old teaching had not been wholly forgotten. Guy de Chauliac,[17] his pupil, followed Galen's advice to apply the ligature to the central end of the vessel when other means failed, a custom which most of the Italian surgeons followed. Silk was the material used by de Chauliac.

With the study of anatomy, which was revived towards the close of the middle ages, surgery made its first modern advances; but the ligature was still chiefly confined to cases of injury where large vessels were implicated. Among the principal changes introduced at this time was the transfixion of the vessel by a needle armed with a double ligature by Bertaplagia,[18] a Paduan professor, who flourished about the middle of the fifteenth century. Giovanni Vigo has been accredited by some with the discovery of acupressure on the strength of the following passage: "modus ligationis aliquando efficitur intromittendo acum sub vena desuper filum strigendo." A process somewhat similar to acupressure we first find mentioned about this time by Mariano Santo,[21] a traveling Italian surgeon, who passed a deep stitch through the flap of a wound including the end of the vessel. In Naples, Alfonzo Ferri[22] used a sickle-shaped needle, with blunt edges, armed with a double ligature; Angelo Bolognini,[23] founder of the school of Bologna, used silk ligatures, but only to a limited extent. The cautery and styptics still retained their hold as popular methods, and were largely employed. After the impetus given to surgery in France by the arrival of Lanfranchi in Paris in 1295, and by his pupil Guy de Chauliac, little progress was made in that country. Among those who were the immediate predecessors of Paré were Jacques Houllier, who used both the hook and the forceps, but timidly; and Jean Tagault, who was professor both in Paris and Bologna, and followed the teachings of Chauliac. In Germany at this period the ligature was known to Braunschweig, the "Senior of German surgery," to Gersdorf and to Ryff.

When Ambroise Paré[24] began to practice surgery the ligature had been in use for nearly two thousand years, but its range in surgery, it will be seen, was extremely limited. Large vessels were tied only when severe injuries forced the surgeon to resort to the ligature, and in surgical operations it was confined chiefly to vessels of moderate calibre. The few instances recorded by the ancients of its use in amputations appear to have been forgotten, and the primitive methods employed at that time for removing limbs have been justly declared to be more worthy of the butcher than of the surgeon. This great surgeon is, therefore, entitled to the credit of not only appreciating fully this method of arresting hemorrhage, but of making the ligature universally applicable. To him this contribution to surgery, which occurred in 1552, naturally appeared in the light of a discovery original with himself, for, with isolated exceptions, the fashion of the day undoubtedly differed little from that in vogue a thousand years before. In the English translation of his works occurs the following passage: "Therefore, I would earnestly entreat all chirurgeons, that leaving this old and too, too cruel way of healing, they would embrace this new, which I think was taught me by the special favor of the sacred Deity: for I learnt it not of my masters, nor of any other: neither have I at any time found it used by any: only I have read in Galen that there was no speedier remedy for staunching of blood than to bind the vessels toward their roots; to wit, the Liver and the Heart." Paré did not attempt to isolate the vessel, but resorted to the mediate ligature. He says: "Il te ne faut estre trop curieux de ne pinser seulement que les dits vaisseaux: pource qu'il n'y a danger de prendre avec eux quelque portion de la chair des muscles, ou autres parties: car de ce ne peut aduenir aucun accident: ainsi avec ce l'union des vaisseaux se fera mieux et plus surement, que s'il n'y avoit seulement que le corps des dits vaisseaux compris en la ligature. Ainsi tirés on les doit bien lier avec bon fil qui soit en double." He sometimes passed the ligature around the vessel by means of a needle, and he also used the forceps. Occasionally the needle and thread were passed through the flap, the skin being protected by a compress, over which

the knot was tied.   He used both single and double ligatures: his
successors generally used several.   He adopted the view of Galen,
that the granulations closed the mouth of the vessel.

In spite of the support given to this method by so great an authority, the example of Paré found but few imitators.   Even Guillemeau,[28]
who was the champion of his friend and teacher, and was sufficiently interested in the operation to invent a new forceps, confined
the use of the ligature to primary amputations, and used cautery in
case of gangrene.

Harvey's discovery, which occurred in 1619, gave but a feeble
impetus to the ligature.   Richard Wiseman,[28] who has been called
the "Father of English Surgery," and the "English Paré," made
experiments for staunching the blood of arteries and veins.   He
preferred the use of a "royal styptic," or the cautery.   Cooke, of
Warwick, refers in 1675 to Paré for a description of the method of
"stitching" the vessels; and adds that it "is almost wholly re-
jected."   Fabricius Hildanus,[26] in Germany, favored the ligature,
but after him there was a decadence of surgery in that country,
and the ligature was rarely used.   Cornelius Van Soligen[30] modified
the forceps so that the jaws remained closed after seizing the artery;
but although he and his colleagues were familiar with the ligature,
and performed Antyllus's operation for aneurism, there was little
progress to record.   Fallopio, in 1660, described the operation
accurately.   He recommended that the nerves and arteries should be
carefully separated with the finger nail, and asserted that the sheath
need not be opened, as it causes but little pain to include it in the
ligature.   He observed the return of the circulation in a limb one
year after ligature of the .popliteal, but thought it took place in the
ligatured vessel.   Marcus Aurelius Severinus (1580-1656) was the
first to tie the femoral artery at Poupart's ligament.   Gaspare Taga-
liacozzi (Taliacotius) still depended, in his rhinoplastic operations,
however, upon cautery and styptics.

In spite of the invention of the tourniquet by Morel[29] in 1674,
and of the aneurism needle by Barthelemy Saviard[31] (1656-1704),
(a blunt needle, "fait exprès pour faire la ligature des artères"),
the imperfect knowledge of anatomy and of the physiology of the
circulation prevented surgeons of that day from appreciating the
advantages of the ligature, and, at the opening of the eighteenth
century, the actual cautery was still the customary method of arresting hemorrhage at the Hotel Dieu.   The contrast between the ligature and the cautery was not found to be so great as might have

been supposed. A glance at Paré's plates shows the forceps, an instrument of rude and clumsy pattern; and it is not surprising to learn that the new method was dreaded by some more than the cautery. No attempt was made to isolate the vessel, but veins, nerves, and arteries were indiscriminately bound together by the "mediate" method. No wonder that patients complained of great pain, with cramps and twitchings, in the stump. And even Petit, with whom modern investigations on the healing of arteries may be said to have begun, (1731), actually proposed compression as a substitute for the ligature.

Sharp,[33] in his "Critical Enquiry into the Present State of Surgery in England" about this time, states that the ligature was used sparingly from a "horrid apprehension of compressing the nerves." But Alexander Munro[34] soon showed the advantages of the direct ligature, and that it could be applied without danger to the vessel, as did also Pierre Dionis[32] in France, at the beginning of the eighteenth century. The latter used chiefly the ligature "en masse," and introduced a sliding forceps, the "Valet à Patin," but he only tied the central end of a vessel.

Petit[35] first called attention to the agency of the thrombus in checking hemorrhage; the blood effused around the end of the vessel he termed the "couvercle:" that found within the lumen he named the "bouchon." Its protective action he showed was only provisional, but it remained for some time closely adherent to the internal wall, and eventually disappeared when permanent cicatrization took place. A few years later, Morand[37] confirmed these views upon the formation of the thrombus, but dwelt particularly upon the retraction and contraction of the divided walls; he also showed that the inner walls were ruptured by the ligature and inverted; and that the thrombus was thus enabled to withstand the blood pressure. It was probably only in exceptional cases that either factor could operate without the aid of the other in preventing further hemorrhage. He attributed therefore more importance to the ligature than did Petit. Pouteau[41] of Lyons, in 1760, denied the constant presence of a coagulum, and maintained that the retraction of the artery had not yet been satisfactorily demonstrated. He thought that the clot was only a weak and subsidiary means, and that the swelling of the cellular tissue around the extremity of the vessel offered the chief obstruction to the bleeding. He therefore advised that as much tissue as possible should be included in the ligature: but the pain caused by this clumsy method made the actual cautery seem mild by comparison. A general protest was soon raised by French, English, and German surgeons against this method, and it was shown that it might even prove a source of secondary hemorrhage. Pouteau, in describing the thrombus, says, " Ce corps conique degorgé dans de l'eau claire, parut très distinctement un sac quasi membraneux en forme d'entonnoir lorgne, rempli d'un caillot de sang noirâtre: il laissoit voir à la loupe une grande quantité de bourgeons, semblable à ceux qui naissent d'une plaie qui commence à s'incarner." It is not probable that these were true granulations, as this was but the third day of the healing process.

In England little value was attributed to the efficacy of the thrombus. Kirkland,[42] with whom other English writers agreed, regarded the contraction of the arteries rather than the coagulation of the blood, as nature's means of arresting bleeding. He cites those cases where a sudden stoppage of hemorrhage follows fainting, and the action of the umbilical vessels in the newly born infant. The vessel is closely contracted for an inch or more from its extremity, the sides become adherent, and it subsequently shrinks into a cord. Gooch,[46] who was the first apparently to expound these views, points out that, after contraction and retraction, the walls of the vessels coalesce as far as the first branch, and their mouths are closed by a growth of tissue aided by the vasa vasorum. White[43] even maintained that the clot prevented closure of the artery, and that it should be removed. In John Bell's opinion the pressure exerted by the surrounding tissue infiltrated with blood was sufficient to prevent hemorrhage, adhesive inflammation subsequently uniting the vessel walls. Hunter[49] gave an impetus to the study of this question by his novel views on the process of healing by first intention. Blood being effused between the lips of the wound, the red corpuscles were absorbed, and the coagulated lymph was transformed into the new cicatricial tissue. He found that spaces could be filled with an injection mass coming from the lumen of the vessel, and thought that, like similar appearances in inflamed surfaces, and the vascular loops in the embryo of the chick, they were the beginning of a new vascular development. It is evident that he believed in the so-called " organization of the thrombus," but he also believed in the agency of the vessel-walls, for he says: "Arteries unite by adhesion when their sides are compressed: this we find after the division of the larger arteries after amputation." The effects of rupturing an artery by tearing or pulling were recognized by him. " It is the property of flexible bodies to have their diameters contracted as they are lengthened: in arteries this might be carried to a great degree when permanent effects are to be produced. * * * Surgeons do not take advantage of this, but farriers and gelders do." He thought that arteries of considerable size were sometimes regenerated. A portion of the carotid artery of a young ram being removed, the circulation was re-established in twenty-seven days; and in another case in eight weeks. He explained the regenerative process in this way—small vessels are seen in the coagulated lymph thrown out, and connecting the ends of the vessel: one of these enlarges and the others atrophy, and the circulation in the vessel is thus restored.

Closely following upon Hunter's came the observations of Jones,[5] whose extensive experiments and thorough and painstaking work have made him the chief authority upon this subject previous to the era of histological research. To this author is due the credit of clearing up the existing confusion. He attributed to the different structures concerned in the healing process their relative importance. His descriptions are so much more true to nature than those commonly found in text books of the present day that they deserve more than a passing reference. His first series of experiments, chiefly on horses, was undertaken to illustrate nature's means of arresting hemorrhage. An artery was divided, and the external wound quickly closed so as to make the division as nearly subcutaneous as possible. The artery was found retracted in its sheath and slighty contracted at its extremity. A coagulum was found within the sheath and external to the vessel, appearing like a continuation of the artery; and, later, a slender conical coagulum within the vessel only partially adherent. Permanent occlusion, he says, was effected by the inflammation of the walls, the vasa vasorum, pouring out lymph which collects between the two coagula, is somewhat intermingled with them, and is firmly united all round to the internal coat. The external clot is soon absorbed, as is also, later, the coagulated lymph, which has produced a thickened and almost cartilaginous appearance in the parts around the vessel. In the meantime the vessel is contracted up to the first branch: its cavity is obliterated, and its condensed tunics are reduced to a delicate ligament.

In ordinary accidents, the internal coagulum contributes to the suppression of hemorrhage, because its formation is uncertain, and, when formed, it rarely fills the canal, and is not adherent unless the walls have been lacerated considerably; and then it adheres to the lacerated spots first, and extends a long distance up the canal of the vessel.

If an artery has been partially divided, there is a coagulum between the vessel and its sheath, thickest at the point of injury. If the vessel is small, the wound is closed with coagulated lymph, which surrounds it in a tumor-like mass. Wounds less than one fourth the circumference of an artery in extent are capable of healing so as to occasion little or no obstruction in the canal. The fact is noticed by Jones and other observers, that it is extremely difficult to produce aneurism in dogs and horses by wounding arteries.

If an artery is surrounded by a tight ligature, its middle and internal coats will be as completely divided as by a knife; the external coat remaining entire. The strength of an artery depends chiefly on its external coat, which answers, in some respects, the purpose of a strong fascia. If no branch be near, a coagulum is formed, which is of little consequence, for there is an effusion of lymph which unites the wounded surface and forms an external thickening, which, involving the surrounding parts, covers over the vessel and forms the surface of the wound. The ligature ulcerates through the external coat, and the vessel is obliterated up to the collateral branches on both sides. Jones was the first to bring to prominence the agency of the walls of the vessel in the process of repair. The organizable lymph, he thought, was poured out from the vasa vasorum, repair beginning in this material, and not in the coagulated lymph of the clot of Hunter.

The researches of Hunter and Jones gave renewed interest to this study, and, during the next three decades, we find a large number of communications upon this subject. In France, Ribes,[66] Bouillaud, Roche, Sanson, Blandin,[81] and Gendrin[71] pronounced in favor of the organizing power of the thrombus. The latter writer says: "If the blood is enclosed between two ligatures in an artery or vein, it coagulates, the serum is absorbed, and a slight inflammation of the walls takes place. The clot decolorizes, a thin layer of coagulable fluid spreads itself over the inner wall of the vessel, and aids in the cohesion of the thrombus, which finally grows to the wall and becomes organized." Henry Lee, repeating Gendrin's experiments, concluded that the clot acts as a foreign body, setting up an inflammation which begins in the outer coat, and extends to the inner coat of the vessel; a natural process for the elimination of the thrombus. He concluded also that an effusion of lymph from the living membrane of the vessel does not take place. Andral,[82] although he accepted the popular theory, believed also, with Cruveilhier,[97] in an adhesive inflammation of the walls of the vessel.

Among German writers may be mentioned Meckel, who was among the first to call attention to the white corpuscle, though ascribing to it no special rôle, and Phillip v. Walther, who accepted the views of Hunter; the former, however, believing that the vessel-walls combined with the clot to form a cicatricial tissue. On the other hand, Ebel denied the participation of the internal coagulum, as did also Chelius, and Rust, as the result of their experience in the healing of wounds; the latter maintaining "that blood, as such, does not become organized; neither vessel nor injection mass are found to penetrate it." In the later editions of their works they gave their adherence to the views of Stilling,[98] whose experimental researches, the most extensive since Jones, seem to have definitely settled the question at that time that the thrombus did become organized. The results of Stilling's investigations coincided with those of Jones, with this exception. The former divided the process of repair of the thrombus into three periods:—at first soft and succulent, the clot grows denser, at the same time diminishing in size: it then becomes vascularized. He regards the vascular spaces as sinuses, and compares the tissue of the thrombus to that of the placenta. The third period includes a development of a fine connective-tissue, which becomes completely incorporated with the vessel-wall, and is followed by a retrograde metamorphosis, which results in an absorption of the vessel and thrombus up to the first collateral branch. The tortuous vessels seen subsequently at the end of the vessel-stump, he suggests, are those developed by the thrombus.

In Italy we find Scarpa, in 1817, maintaining the adhesion of the walls of the vessel without the intervention of the clot.

Guthrie [80] called attention to a fact not to be forgotten in drawing conclusions from experiments on animals, that the repair of their vessels does not follow the same course as in the case of man. His account of the process is exceedingly accurate. In dogs, if an artery he wounded through more than three quarters of its circumference, a coagulum forms and extends up and down the sheath, the cut edges and the surrounding parts inflame, and throw out lymph, which adheres to and surrounds the clot. The lymph becomes organized, and unites the edges so as to leave no mark. The cicatrix may, however, yield, and, as in man, an aneurism will be formed. In man, a longitudinal slit may heal without obliteration of the canal in medium-sized arteries, as the temporal: in larger vessels the

canal becomes obliterated. If completely divided, the vessel contracts at its end at first, and later, at about an inch from the end,
this section being filled with a coagulum. Still later the contraction
takes place up to the first large branch, so that, out of four or five
inches, two or three inches will be impervious, the remaining part
being very much narrowed. Vessels of considerable size are capable
of arresting hemorrhage without aid, the power and influence of the
heart over the circulation through the arteries being greatly overrated.

Amussat [107] also calls attention to the peculiarities of the healing
process in animals. He states that aneurism can never be produced
in dogs, and in horses only the arterio-venous variety is observed.
He also thinks that the reparative process is more independent of
the ligature than was supposed at that time. Many vessels would
cicatrize, if time and rest were given them. More reliance should
be placed on pressure. Many other names might be mentioned,
but the authorities quoted serve to indicate the prevailing opinion
of that period.

With the rise and progress of histological research, the question
of the organization of the thrombus was subjected to new tests.
Schwann, in 1838, had developed his " Cell theory," according to
which all tissues are developed from primary cells. A direct change
from amorphous blastema could therefore no longer be maintained,
cells forming first in the cytoblastema or fibrine, from which the new
tissue is developed.   Henle, however, adhered to the old view that
the coagulated fibrine was immediately changed into connective tissue, and it was under his direction that Zwicky [110] repeated the experiments of Jones and Stilling, adding an examination of the
thrombi under high powers of the microscope. This was the first
elaborate microscopical study of the question. The special point
to be determined was the appearances to be noted in the development of fibrine mingled with blood corpuscles; the " reorganization
of the blood in substance " was not doubted by Henle. The following changes were noted in the thrombus. At about the fifth day
certain granular bodies appear, which, it is assumed, are developed
directly from the fibrine; these do not appear to have any special
significance, for, before the end of the second week, they have broken
down.   By the eleventh day the fibrine has lost its fibrillated appearance, and has become a homogeneous structureless cytoblastema.
At the beginning of the third week this mass begins to have a
slightly fibrillate appearance longitudinally, and, at the sides, band-
like fibres are noticed. Numerous round and oval granules appear

on the addition of acetic acid, and near them are red granules, the remains evidently of broken-down blood corpuscles. The broad fibres increase in number until the whole mass of fibrine has been thus transformed. They appear, says the author, uncommonly like muscular fibres, but later on, about the sixth or seventh week, these fibres become still further broken up into wavy fibrillæ, corresponding to those seen in connective-tissue. The red blood corpuscles take no part in the tissue-formation. Zwicky, like Stilling, observed the formation of vessels in the thrombus, and he agreed with the latter that it became organized; but he did not find the subsequent absorption of the cicatrized thrombus and vessel mentioned by that investigator. This was in direct opposition to the views of Rokitansky,[106] whose book appeared the previous year, in giving prominence to the action of the vessel-walls, and pointing out that the obliteration of the portal vessels took place without the intervention of a thrombus. Rokitansky says: "We believe that the closure of a ligatured vessel takes place without the aid of the thrombus, the inner coagulum, that this is an accidental formation, and by no means a condition necessary for the obliterating process. We are of the opinion that the closure and obliteration of the ligatured vessel is essentially the same process which occurs in vessels which have been excluded from the active circulation, and, owing to the change in the blood current, have become useless; for example, the umbilical arteries and the ductus Botalli. The end of the ligatured artery having become closed by a fusion of the inner walls thus brought into contact, the further obliteration takes place by a gradual diversion of the current into collateral channels and a corresponding narrowing of the cavity of the vessel until the inner wall either coalesces, or a new layer deposited upon it completely closes the lumen. The so-called vascularization of the thrombus we have never observed, and believe this condition to be the same thing as that remarkable appearance which we have recognized as canaliculization." These opinions, it should be remembered, were expressed at a time when the views of Hunter, Schwann, and Henle had been generally adopted by the scientific world.

About this time, the work of Porta[109] appeared, a contribution to certain branches of this subject too elaborate and too excellent to be passed by unnoticed. He made a special study of the collateral circulation, and of the kind of material to be used as ligature.

Virchow,[108] at first, was more or less influenced by Hunter's views on the organizing power of coagulated fibrine; when, however, he became convinced of the power of cell-action, and inaugurated the modern system of cellular pathology with the formula, "Omnis cellula e cellula," he devoted his attention to the action of the white corpuscles, found in the thrombus, particularly in his studies of thrombosis and embolism. He was fully convinced of the organization of the thrombus; there was only a question in his mind as to the proper interpretation of the process. He says: "There is certainly no doubt that organized tissue forms in those places where fibrine or blood clot previously existed, and that it would not have been developed, had not the clot been there." Although he conceded that the vessels of the thrombus might grow from the vasa vasorum, nevertheless, he did not think that the presence of the cellular element of the new tissue could be explained by a primary growth inwards of the cells from the vessel-wall. In studying the contraction of the thrombus, in 1850, he noticed that the white corpuscles became elongated, egg-shaped, and spindle-shaped; that also, when compressed in the meshes of the shrinking clot, the cells became extremely elongated or stellate, so as to look not unlike bone-tissue; the nuclei also became elongated. "It would be impossible to believe," he says, "that these stellate cells are originally white corpuscles, unless one had seen, as here, side by side, all stages of development." At that time he thought the cells merely passive in their changes. In a second series of experiments, undertaken the following year, Virchow studied the earliest changes, as seen in the membrane formed over the pyramidal plugs of rubber introduced into the pulmonary arteries of dogs. On the seventh day, the fibrine was homogeneous, and contained stellate cells, partly anastomosing with one another, and partly isolated; the appearance being, as before, not unlike ossifying tissue. As early as the second day similar changes were noticed, and he was forced to fall back upon the white corpuscle seen in the fibrine for an explanation of these phenomena; and he says, in conclusion, "Must we not concede that white corpuscles may be the originators of the future connective tissue corpuscle?" He described the "sinus-like degeneration" as the hollowing out of channels in the thrombus by the blood-current, and distinguished it carefully from the vascularization of the thrombus.

In the same year appeared a communication from Reinhardt [112] particularly worthy of mention for the correctness of certain views which hold good to-day. He disproves the direct organization of fibrinous exudations, deriving young tissue, which forms on its

site, or that of a blood clot, from the surrounding tissue; the formative material exuding from the blood-vessel being transformed,
according to the Schwann theory, into cells capable of development.
"As the new tissue grows into the fibrine the latter is absorbed.
There is no trace of new formation of nuclei in the fibrine, the development of the young tissue takes place entirely independently of
the fibrine." Concerning the organization of the thrombus he
states: "Around the ligature the tissue becomes inflamed, and in
consequence of the suppuration which takes place, granulations form,
which mingle with the cellular tissue, and take the place of the ne-
crosed portion of the vessel, and grow into the thrombus just as
new tissue grows into the fibrine in the healing of wounds." He
never saw any such changes in the fibrine as were observed by Zwicky,
there being no trace of organization there: it was amorphous as at
first, and only occasionally were granular cells to be seen in it, these
being probably pus corpuscles. Notta,[13] in Paris, maintained, in
a thesis published in the same year, that blood clot was capable only
of a retrograde change: he, like Pouteau, thought that cicatrization
began in the surrounding cellular tissue.

Rokitansky, in 1856, yielded so much to the prevailing views as
to allow, in certain cases, a direct organization of the thrombus into
connective-tissue: in other cases, again, it was reabsorbed, and independent vascularization did not take place. The bulk of the
growth, he still maintained, came from the vessel-wall; but he no
longer derived it from the plastic exudation, but from a connective-
tissue growth from the interior. The tide had now fairly turned,
and, under the stimulus of Virchow's teaching, the views of Schwann
and Henle were rapidly disappearing, but his opinion on the organization of the thrombus restrained for some time a farther advance
in this direction.

The work of C. O. Weber,[19] confirming Virchow's views, obtained
general acceptance throughout Germany, and is even to-day an
accepted authority in many text books. The following quotation
expresses his views: " Within the thrombus the organization begins
in a few hours, and is at first entirely accomplished by the so-called
white corpuscles. The red blood-corpuscles take no part in the
process. They soon yield their coloring matter, which is precipi-
tated in the fibrine; they shrink, and break down more or less rapidly.
The fibrine of the blood clot also breaks down into a fine granular
detritus. The striation and stratification of it has no special significance, for the fibrine has as little to do with the changes as the red

2

corpuscles. The colorless corpuscles are here, as elsewhere, specially concerned in the process of organization. In the first hours
after the formation of the clot, these bodies are seen to take on different forms; occasionally they elongate themselves into a spindle-
shape, the mass of protoplasm sending out on both sides delicate
prolongations which unite with similar ones from other cells, and,
by disposing themselves side by side, help to form bundles dotted
with nuclear swellings; or, several such prolongations will be thrown
out in different directions, producing a stellate net-work, with nuclear
enlargements, such as are seen in young connective tissue, which
grow into and occupy all parts of the fibrinous clot. Finally, an
enormous multiplication of these bodies takes place; they undergo
segmentation, divide, and increase rapidly; and, where there were
few before we now find large masses of cells. These changes trans-
pire during the first few days after the coagulation of the thrombus.
At the end of a week, canals are already seen running in various
directions through the thrombus filled with rows of red corpuscles."
That this mesh-work is a genuine anastomosing system of blood-
vessels admits of no doubt. This process is the same in the veins
as in the arteries, and the smallest branches of the latter give the
same appearances as those seen in the large trunks. In thrombi
adhering to one side of the vessel-wall, particularly those formed
in the pockets of the valves of veins, the same changes may be observed. In this way the "vascularization" is effected by the white
corpuscles, which are the sole organizing elements in the thrombus,
through the canaliculization of the thrombus which precedes; and
it is finally completed when the young vessels become united with
those of the vessel-wall. The "canaliculization" of the thrombus
is the name given to the formation of the canal, and it precedes the
"vascularization" or the union of the canals with the vasa vasorum.
These blood canals appear to form by the longitudinal grouping of
spindle-cells, or by the dilatation of the anastomosing processes of
young connective tissue cells into a tubular system, and are found,
first in the periphery, and later in the centre of the clot. They
join those of the vessel-wall in the third or fourth week. These
vessels pass into the thrombus at the point of ligature, but sometimes higher up. By the sixtieth day the thrombus is full of vessels,
and there is usually a large one in the centre. Later, the thrombus
shrinks to a small connective tissue plug.

According to Rindfleisch [44] also, the first changes occur in the
white corpuscles; a delicate protoplasmic net-work being formed,

with nuclei at the point of union of the meshes. He likens the clot to a connective substance in which the white corpuscles represent the cells, and the red corpuscles and the fibrine the matrix. The later changes of the organized thrombus he describes as the cavernous metamorphosis.

In spite of the weight of such authorities, there were still those who did not accept the theory of the organization of the thrombus, and, with the investigations on the action of the endothelium, the study of this question entered upon a new phase.

Cohn,[128] in 1860, was the first to suggest these cells as organizing elements, and Lanceraux, speaking of the fate of the embolus, states that it disappeared, as pseudo-membranous formations grew into it from the vessel-wall. Foerster, also, could not convince himself of such an organization of the thrombus, and so states in his text book published at that time. He doubts also whether the white corpuscles undergo any such changes as were described by Virchow. About this time came also an important communication from that eminent histologist His,[143] which did away with one of the strongest objections to the theory of the activity of the lining membrane of the vessel in adhesive inflammation. His showed that the cells of this structure were of an essentially different origin from the epithelium of the skin and mucous membrane, that they arose from the middle germinal membrane, belonged to the group of connective substances, lined the inner wall of the blood and lymphatic vessels, the serous membranes and the surfaces of joints, and gave them the present name "endothelium." Waldeyer,[154] consequently, recognizing that the endothelium was capable of producing connective tissue, and that, by its growth, genuine granulations could be produced, convinced himself that this tunic took an active part by producing a young vascular connective tissue within the vessel. He says: "The so-called organization of the thrombus and of extravasation of blood occurs from the vessel-wall or from the neighborhood of the extravasations; and, in the case of the vessels, the epithelium plays the most important rôle by growing into the thrombus. The intima becomes vascularized through the media, capillary loops shoot into the thrombus, accompanied by delicate bands of spindle-cells, which form the basis of the future connective tissue. The extravasated blood never organizes, but is invariably absorbed, leaving behind more or less coloring matter."

The investigations of Recklinghausen,[136] which led to the discovery of the wandering cells, and the re-discovery by Conheim of

Waller's observations of the passage of the white corpuscles through the walls of blood-vessels gave a new turn to the discussion. Here, then, was a fertile source, whence a supply of cells could be obtained, wherewith to people the thrombus, the study of which, like any other departments of pathology, was strongly influenced by these important contributions to medical science. A pupil of Recklinghausen, Bubnoff,[149] undertook to demonstrate the passage of such cells through the walls of veins which had been tied. After the ligature had been applied, granules of vermilion were rubbed over the vessel externally, and were found to be taken up by the wandering cells, and to be carried into the thrombus through the vessel-wall. Bubnoff concludes, "that the white corpuscles of the thrombus lose their power of wandering, and take no part in the cell-formation of the organizing tissue. Cells take part in the organization of the thrombus, which creep in large numbers into the vein. The mass of the cells are probably derived from the layers of the vessel-walls, and from the surrounding tissue." Billroth[137] adopted these views, but still maintained that proliferation took place in the white corpuscles of the thrombus. While the view of Conheim,[153] as to the origin of cells in inflamed tissues, was adopted by a large number of pathologists, there were those who still believed that the pre-existing cells of the part were capable of proliferation. The most prominent exponent of this school is Stricker, who goes so far as to believe in a subdivision of even the intercellular substance, and its conversion into amœboid material. The artery, with its delicate lining of cells, easily isolated from surrounding tissue, seemed particularly adapted to test the disputed points, and many of the papers subsequently produced were written by adherents of Conheim or of Stricker. Argument and investigation were now employed chiefly to determine the relative merits of the emigration and proliferation theories, both parties conceding that the thrombus in the old sense could not become organized.

Thiersch[159] arrayed himself with those who saw in the new cells a "proliferating vessel-epithelium," and believed that the vascularization of the thrombus was effected partially, if not entirely, by these cells. In sections through the epithelial covering of the follicle, and through a part of the coagulum of a corpus luteum of the sow, he observed newly formed large cell-masses, interspersed with vessels which project in a series of loops into the clot, in the white corpuscles of which no change could be observed. This explained Bubnoff's results with vermilion granules by assuming that the granules were not carried by cells, but floated in the plasmatic current, which flowed through the intercellular spaces of a network which forms around wounded vessels, and were thus carried into the interior of the vessel. Bubnoff repeated the experiments the following year with reference to the objections raised by Thiersch, but still held fast to the view that the wandering cells were the sole transports of the pigment-granules

Kocher[166] and Szuman endeavored to restore the Weber theory of the organization of the thrombus, and they appear to have been the last to uphold it. The work of both writers was largely devoted to the changes found in acupressure. Kocher gave special attention to the action of the endothelium, and convinced himself that it not only took no part in the process, but even hindered it. The vascularization could not, therefore, take place through these cells; and, in further support of this view, he showed that the larger vessels are always in communication with the lumen of the vessel, and are never found to make their way through the uninjured wall. After the double ligature of a vessel, the blood having first been excluded, no union of the walls of the isolated portion takes place. The theory of Deschamp and Lawson Tait,[144] of union by first intention, he does not think tenable. The comparison of the intima with the peritoneum is not justifiable: it would be more correct to liken the adventitia to the peritoneum and its action in suture of the intestine. Moreover, in some cases the intima and media are separated from their corresponding layers, and therefore it is impossible for them to unite by first intention. The union of the adventitia is doubtful, for it is not probable that it would be able to resist the blood pressure. The thrombus is therefore, in Kocher's opinion, the most important factor in the temporary and final suppression of hemorrhage. Even in cases where the thrombus is exceedingly small, owing to the presence of a branch, we find it driven in between the clefts in the vessel-walls; thus affording its protecting influence. The clefts in the walls are not, of themselves, sufficient to check the blood-current; for, if the ligature be immediately removed, the circulation is restored. On the other hand, if these ruptures are not made by the ligature, the blood will remain fluid, and, when released, will mingle again with the current. We must have, therefore, not only obstruction to the circulation, but simultaneous bruising of the vessel-wall, to produce conditions favorable to permanent closure. The old view of Petit, on the efficiency of the inner coagulum, he thinks still holds good. In small vessels temporary pressure may suffice, owing to the rapid re-establishment of the collateral circulation. The pressure exerted by the coagulum of extravasated blood acts in this way. As to the canalization of the thrombus, it is thus explained. According to the manner in which the fibrine is deposited the blood will take certain directions in filtering through the clot, as changes are going on in it, and, on the borders of the canals thus formed, a growth of connective tissue takes place by the development of the white corpuscles into spindle-shaped cells.

Szuman [189] agreed essentially with Weber, but did not accept the latter's theory of the formation of vessels by the tubular metamorphosis of the prolongations of cells. In the same number of the journal, in which Kocher's article appeared, there was one also by Tschausoff, [165] which takes directly opposing ground. This investigation included, in addition to those on animals, several on the human subject, among which appears a specimen taken from the brachial artery of a stump twelve years after amputation. Tschausoff concluded that the thrombus takes no part in the organization, but breaks down in all its elements, as the growth forms from the vessel-walls; the function of the thrombus being merely to offer a temporary protection against hemorrhage. The cicatrization is accomplished by the connective-tissue elements of the vessel-walls. The thrombus and ligatures are the source of irritation which produces this growth. Neither muscular fibres, nor endothelium, take any part in the process: the new vessels form entirely from the vessel-walls. The work of this observer, being a graduation thesis, and but very crudely illustrated, was hardly equal to sustaining so independent a position.

Cornil and Ranvier [184] conclude from their studies that the obliteration of the artery, after ligature, is effected by a new formation, produced by an arteritis, the result of the traumatic lesion springing from the intima. The early changes noted in this coat are swelling of the cells, and multiplication of the nuclei, and a few days later, a thickening of the intima by cells which appear fusiform, but are really flattened. These resemble endothelial cells or connective tissue cells swollen by inflammation. They differ in no way from those seen in acute endarteritis. By the eighth day granulations are formed: by the fourteenth, these contain capillaries. Gradually the outline of the coat is lost, and the capillaries appear to come from the vasa vasorum. As the granulations fuse together, the clot disappears: it does not become organized. The wandering of cells through the walls of vessels after single ligature

does not take place. Bubnoff's experiments still remain the subject
of much controversy. It was conceded by Mayer[177] that the cells
found in the early periods in the thrombus were the product of the
white corpuscles, and the wandering cells; but he also maintained
the growth from the vessel-wall. Durante[180] made the vermilion
experiment and the double ligature the objects of a special study.
He found that the walls of the vessel between the two ligatures
undergo a necrosis, and that the white corpuscles penetrate them,
as they would an inert body. In single ligature he could not find
cells containing granules either in the clot or the walls of the vessel.
If the pigment be rubbed in hard, he has occasionally found it inside
the veins in a free state. He therefore concludes that the wandering cells have nothing to do with the process, and attributes the work
to the endothelium, or the layer of ramified cells found beneath the
endothelium in the intima.

Maunder,[195] in a series of lectures, gives some valuable hints as
to the formation of the blood clot. A ligature, he says, to be
successful, must give rise to an inflammatory process in the shape
of adhesion of the cut surfaces and of the divided coats, and a destructive process in the shape of ulceration, by which the ligature
is to be cut loose. According to Lister and Callender the greater
turmoil in which the blood is thrown on the cardiac side, by imping-
ing against the obstruction, accounts for the more rapid coagulation
on that side of the ligature. This churning process leads to the
deposit of fibrine more or less pure; the deposit of blood clot, on
the other hand, is formed by slowness of movement, as on the distal side in the lower extremity when the anastomosis is not free. In
a case of secondary hemorrhage, after an antiseptic catgut ligature,
it is interesting to note that, subsequently it was found that the inner
coats of the vessel had not been divided, and that there was little
clot on either side.

A decided advance in our knowledge of this process was made
by Baumgarten.[200] As the title of the article indicates, the
author still felt the necessity of combating the old theory, and
he maintains that in all his experiments, which were conducted
antiseptically, no clot was formed. But it was chiefly from
his observations on the growth of granulations from the surrounding external tissues into the interior of the vessel, that his
monograph acquires especial interest. The pressure of the ligature, he says, absorbs the bridge of adventitial tissue, and the surrounding granulation tissue grows in between the ends of the media,

or through other ruptures in the sides of the wall. This tissue is the source of the vascularization of the thrombus, for the pure endothelial growth which also occurs is without vessels. This shows that, in the developed organism, a new formation of blood-vessels occurs only in conjunction with pre-existing blood tubes. Baumgarten's work included, also, a very careful study of the behavior of the endothelial cells. An irritation being produced upon the adventitial and the peri-adventitial tissue, by the ligature, the nutrition of the endothelium, which is dependent upon the vasa vasorum, must be affected; and we find a cloudy swelling of its cells and proliferation. These change also into a cube-shaped endothelium. There soon appears to be a new layer of tissue between the original endothelium and the lamina. It is interesting to note that he mentions, in the outer layer of this new cell-structure, some cells like those described by Heubner in syphilitic arteries as resembling muscular cells, but all this new tissue develops into a connective tissue finally. He thus recognizes, in the new tissue, a central portion composed of granulation tissue containing vessels, and a peripheral non-vascular tissue composed of endothelial cells. He does not believe that the white corpuscles of the thrombus take any part in the new formation. He considers the endothelium the most important factor.

It appears that the cumulative evidence produced at this period was not without its influence upon the standard authorities; for Billroth, who, in accepting the views of Conheim and Recklinghausen on the action of the wandering cell, had said, "After having abandoned the idea of proliferation of stable tissue cells in inflammation, we can no longer talk of the proliferation of intima in the old sense," now suggested to two of his pupils the advisability of going over the ground again; and, as a result, there appeared articles by Raab on the cicatrix following ligature, and by Winiwarter on endarteritis; both of whom endorsed strongly the theory of an active endothelium growth. Raab [28] placed double ligatures on the walls of veins and arteries in such a way as to exclude the blood clot, and to avoid injury, as much as possible, to the vasa vasorum by stripping up the sheath. Cross sections of arteries prepared in this way on the twelfth day show a marked change in the endothelium, oblong or spindle-shaped cells filling out the folds in the lamina with an abundant intercellular substance, and, on the surface, presenting a normal endothelial covering. This tissue becomes fibrous later, but never vascular except near the ligatures. In veins, at the same period, we find a proliferation in the outer coats, and, as there is no lamina to

oppose a barrier to these cells, they become mingled with the growing endothelium. The new tissue in this case is, therefore, more
like granulation-tissue. The earliest changes seen in endothelial
cells is best observed in cross sections of small vessels. An en-
largement of the nuclei occurs, which pushes out the cells more
prominently into the lumen. The old nucleus disappears finally,
several new ones take its place, and an active proliferation becomes
apparent. In veins, the round cells prevail as a type; in the arteries,
flat and spindle-shaped cells prevail. These cells get their nutriment from the vasa vasorum, and, if we wish to study the changes
in them, the vasa vasorum must not be destroyed. Raab calls particular attention to the ramification of spindle-cells from the intima
into the clot, seen well at the fourth day. Some of these anastomosing structures are described as protoplasmic masses, in which, later,
nuclei are developed, either by a differentiation of the material, or
by their being carried to a certain point by a fluid stream in the mass.
Later, these cells collect into bundles covered with endothelial cells,
and, anastomosing with one another, leave spaces like those described
by Weber, and incorrectly supposed to be newly formed vessels.
Still later, they assume the appearance of granulations. Vessels
come from the walls, finally, through the injured spots near the ligature. There is, also, at those points, some growth of a connective-
tissue into the vessel. Dudukaloff ascribes the chief work to this
granulation tissue; according to Raab, it is sometimes the endothelial and sometimes the granulation-cells which predominate, probably according to the injury done by the ligature. In the most favorable cases we may have union by first intention through a growth
of the endothelium solely. The organized tissue consisting first of
cells; later, of cells, fibres, and vessels. Gradually, with increasing
age, the cells disappear, as do also the vessels, and a fibrous band
remains as the permanent cicatrix.

The old view that a thrombus must reach as far as the first collateral branch, is exploded, both through experiments on animals
and through observation on man. There is no way of controlling
its growth. As to the question of the organization of the thrombus,
Raab denies participation of the included white corpuscles; as white
corpuscles may be produced from endothelium, it is not necessary
to account for their increased numbers by supposing an immigration
of wandering cells. He could not reproduce Bubnoff's experiments.
As other cells may take up pigment granules, their presence in the
new tissue does not afford evidence of a participation of the white

corpuscles. Wandering cells may be found in the thrombi of veins, but their action there is doubtful. The growth of endothelium and of granulations is as easily demonstrated as any fact in pathological histology, but the question of the action of white corpuscles and wandering cells is founded on the doubtful basis of hypothesis. A portion of the clot is taken up by the current and carried into the circulation, another portion is consumed and assimilated by the young tissue, till finally only a few hematoidin crystals or amorphous pigment granules bear witness to the presence of blood.

The study of idiopathic endarteritis has contributed much to the belief that the endothelium plays an important rôle in the pathological changes in disease of arteries. According to Cornil and Ranvier, in chronic endarteritis in vessels of small calibre, such as are seen at the base of the brain, the internal tunic has vegetated in such a way as to obliterate completely the vessel; and in the tissue thus formed, we may have a new formation of blood-vessels. In the endarteritis, which precedes aneurism, a similar growth is noticed, which may either narrow the vessel or project into its lumen in granulation-like masses. It is worthy of notice that, in the dilatation of aneurism, as described by these authors, rupture of the media takes place, and the adventitia and the intima become united, vessels growing into the latter from the former; conditions closely resembling those produced by the ligature. Heubner, in studying the changes in the cerebral artery in syphilis, observed a growth in the deep layer of the intima. In the space between the endothelium and the lamina there is a single layer of nuclei, imbedded in a granular cloudy substance. Protoplasm formed about these cells, and new cells were also produced by a proliferation of the endothelial cells, which only grew outward into this layer, and not into the lumen. They eventually formed a new lamina; in fact, a new membrane, similar to the normal membrane, was produced. Friedlander, in describing such a growth between the lamina and the endothelial layer, in obliterating arteritis, suggests that the new cell-growth may come from cells which emigrated from the vasa vasorum, and that the endothelium, and also the white corpuscles of the blood, may participate. Winiwarter's [97] observations were made on a single case, but they are given as an example of the minute changes observed in this disease. In the small arteries, the lumen was nearly filled with a new cell-growth; in the centre was an irregularly shaped lumen surrounded by endothelial-like cells, outside of which were circularly arranged spindle-shaped cells, and fibrous tissue giving the impression of a newly formed vessel. Sometimes two such vessels were seen. The muscular walls were thickened by an increase partly of the muscular cells, and partly by cell-infiltration occurring secondarily, by an invasion from the intima; or we may have the lumen completely filled with a concentrically arranged connective-tissue, the lamina disappearing entirely. In the tibial artery the intima was greatly thickened, and Winiwarter found in the new growth several white concentric circular bands, which he calls newly formed elastic laminæ, between which spindle and irregular shaped cells were found; further in was a myxomatous tissue, and if the blood clot was adherent, these cells grew into it, or a new endothelial membrane was formed by short spindle-cells closely packed together. A number of arterioles were seen, giving the tissue a cavernous appearance. In some places the lamina was found to become granular, then fibrillated, and then to break up and be reflected backwards, permitting a growth of the cells into the media. In some cases a new lamina is found just beneath the endothelium, outside of this being a concentric layer of fibres, giving the appearance of a new vessel-wall formed inside the old one.

In the veins there is a considerable thickening of the intima, generally on one side, complete obliteration only taking place by the formation of a thrombus. The inner layer of this growth is composed of a more abundant mucous tissue than in the arteries, and is covered by a growth of round cells, and not covered, as in the arteries, by an endothelium; consequently, coagulation easily takes place. Outside of this layer, and close to the lamina, is a layer of spindle-cells closely resembling circular muscular fibres. The growth described in these vessels arises, according to the author, from the intima, and chiefly from the endothelium, the media and adventitia playing a subordinate rôle. The finally obliterating tissue is a fibrous tissue rich in cells, and the subsequent formation of blood-spaces is purely mechanical, being caused by the blood-current forcing its way through the tender tissue. They may, in their turn, be obliterated by a further growth, there being a constant struggle, as it were, between the blood current on one side, and the new growth on the other. The new laminæ are probably formed from endothelial cells, which are closely pressed together, become flat and stationary, and lose their nuclei. The muscular cells are found only in the veins, which have, in the lower extremity, such cells normally in the intima.

The smaller vessels become more quickly obliterated than the

larger, being converted into a fibrous cord. The new vessels form
a sort of collateral circulation.

Wyeth [230] finds, in traumatic arteritis from external injury, a
hyperemia with cell-emigration and proliferation in the arterial coat,
the connective tissue cells of the adventitia, the white corpuscles,
and the endothelium all taking part. The vessel is narrowed by a
growth from the intima, and capillaries finally find their way into
this new tissue. In idiopathic endarteritis the growth occurs pri-
marily in the intima, the outer coat being affected secondarily. In
syphilitic arteritis Wyeth observed an irregular thickening of the
external coat, but he found the chief change in the intima; although,
in two cases observed by Greenfield, all coats suffered some change.
In his illustrations some spindle-shaped cells are seen, and also in
one case, a newly formed elastic lamina.

It will be seen that the weight of evidence appears to be strongly
in favor of a growth of the cells of the intima, and, at the present
day, this seems to be the generally accepted view; although a few
still hold to a participation of the white corpuscles to a certain degree. For instance, Lee and Beale [155] made investigations on the
arteries of horses and of a donkey, to show that repair did not take
place by means of adhesive inflammation of the internal wall of the
vessel. After describing the external clot they state, " The opening
in the elastic coat of the vessel is not, however, immediately occupied by blood, but, as has been shown, is gradually filled up with a
perfectly colorless substance. This material, which resembles the
fibrine found in some aneurismal sacs, is, like that substance, deposited from the blood, layer after layer, until the space is filled up.
Sometimes the process continues until an actual elevation, projecting
above the level of the inner surface of the artery, is formed." Of
the laminæ of transparent fibrine they say, " It is evident they have
been formed from the blood which flowed along the vessel, and not
from any material poured out from beneath by the vasa vasorum
or from arterial tissue. We do not, however, regard the material as
a new deposit of fibrine from the blood, but are disposed to think
that it is formed by the agency of the white blood corpuscles which
we know would adhere to the surface of the blood-clot which occupies
the lower part of the wound. It seems probable that these masses
of germinal matter, as they slowly move over the surface, form the
material allied to fibrine figured in the drawings." Later changes
than those seen at three days were not studied, but the authors leave
it to be inferred that, in their opinion, it is probable that a connec-

tive tissue cicatrix may be formed from the material described, although they do not deny that it may be derived from the cells of the arterial wall, "which may gradually increase in number near the cut surface of the vessel, and thus extend from this point into the temporary texture occupying the wound."

This line of investigation was continued by Schultz,[20] who maintains that the process of repair is completed exclusively by the white corpuscles in the vessel. The gaping wound, made by a longitudinal incision, is filled by a clump of white corpuscles, which become fused together (white thrombi) into a homogeneous fibrinous mass, the corpuscles ceasing to perform any further act as soon as they are caught in the clot. Subsequently, this mass is seamed with a series of canal-like spaces ("canalized fibrine"). This fibrine forms a sort of sac at the mouth of the wound. The canals are filled subsequently with more white corpuscles, which multiply rapidly until the fibrine disappears, and granulations are found in its place, between which, as the wound is filled, spaces are left which gradually are formed into small vessels. The cells on the outside of the vessel, being shut out by the fibrine layer, have nothing to do with the process. The sac in aneurism is formed from such a layer of fibrine. Experimenting with the double ligature, Schultz found no sign of action in the endothelium. An important and original observation is made in this paper upon the eventual shape of the cicatrix. If there be no branch near, the cicatricial tissue will be symmetrically developed in a crescent shape; but should such a branch exist, the new tissue will be built up on the opposite side. If several branches are given off, one beyond the other, there will be a thickening of the new tissue opposite each, and the cicatrix may be continued in this shape some distance up the side of the vessel (see Thoma).

Schultz's paper brought out a reply from Pfitzer,[21] according to whom such wounds are filled with the white thrombus, inside of which a growth of endothelium takes place over the wound, and a granulation tissue is formed at the same time outside of the vessel at the same point; these two growths finally displacing the clot between them. The walls of the vessel remain gaping as on the first day, and are held by the connective tissue cicatrix. These are practically the views of Baumgarten, to whom Senftleben[22] replied in the same number of the Archives. A curious and interesting series of experiments was made with the double ligature to prove the inactivity of the endothelium and the part played by the wandering cells. In the usual experiment with the double ligature he found no change in the walls of the vessel, except close to the thread, where there was a slight spindle-cell growth closing the vessel inside, and also a slight infiltration of the media and of the adventitia. In case of suppuration some pus-cells were also found inside; if no suppuration occurred, only a few epithelioid (granulation) cells and giant-cells were found in the blood which flowed out on opening the vessel. When a saline solution, with vermilion in suspension, was introduced into the circulation, pigment particles were afterwards found in the cells inside the segment, showing that they were cells wandering from the small vessels into this space. This segment was eventually converted into a solid cord by a growth penetrating at the ligatured point. In one case a double-ligatured segment was filled with alcohol so as to destroy the lining endothelium, and yet the process was in no way changed. In another series of experiments the double-ligatured portion was immediately removed, and, having been soaked for two days in alcohol, was placed in the peritoneal cavity of rabbits. The same form of spindle-cell growth was found inside these spaces as that described by Raab as coming from the endothelium, although here those cells were manifestly destroyed. Wandering cells were also found in the walls. The swollen endothelium described by Raab, he thinks, was probably attached white corpuscles. If such views were correct, then we should have all small vessels, in an inflamed part, obliterated. The endothelium may be stimulated into reproduction by inflammation, but this limits itself to a regenerative process. The new endothelium is, however, formed from the vessels in the neighborhood, the ligatured vessel being converted into a cord. The arteritis produced by the ligature differs in no way from other forms of obliterating arteritis. Even in the syphilitic form, the walls are infiltrated with wandering cells. Baumgarten, in answer, says that there may be white corpuscles in the new tissue, but that they have nothing to do with the healing process. The genuine formative cells are the so-called epithelioid cells (large cells with one or more nuclei), and these, he claims, are absolutely identical in form with the first descendants of the growing vessel-endothelium. In repeating Senftleben's experiments, he found no wandering through the walls of pieces of vessel placed in the abdominal cavity, but if left there long enough, the ends might be absorbed, and granulation tissue could then grow in. Senftleben used also pieces of lung, and saw the immigration of wandering cells, and their change to spindle-cells; in the centre were broken down white corpuscles. Baumgarten saw in this fatty degeneration an evidence that the white corpuscles are of no importance. The spindle-cells found there are the result of an ingrowth of granulation tissue.

Shakespeare [213] derives the new formation from the endothelium and the subjacent cellular elements of the intima. Already at the end of twenty-four hours a collection of these cells is seen at the point of ligature, forming a cushion on which the clot is seen resting. The latter this observer styles the fibrinous clot, and the former the plastic clot. The plastic clot is composed of cells of a great variety of shapes, and it continues to grow, pushing up the fibrinous clot. On the sides, the growth of the cellular covering of the intima sometimes extends as far as the first collateral branch. It begins to show signs of vascularization as early as the sixth day; later a rich capillary net-work is seen in communication with the open lumen. A communication is effected with the vessels of the walls of the artery between the fifteenth and thirtieth day, at the bottom of the arterial stump, where the intima and media have been cut through by the ligature. The plastic clot cicatrizes, undergoes cavernous transformation, and finally disappears, the only remains of the vessel and the clot being a tough fibrous cord. If pressure be exerted upon the wall of the vessel just above the point of ligature, sufficient to bruise the inner wall, there is an accumulation of cells in the intima and inner layers of the media, the plastic clot forming mainly at that point instead of at the level of the ligature. By compressing an artery for a few hours with a pair of forceps or a serre-fine, an inflammation may be excited by which the lumen may become obliterated. This method is suggested in cases of atheroma or aneurism. The author has actually observed the separation of an endothelial cell from the wall of a capillary in the frog's mesentery, during inflammation. He concludes that the endothelium may be the origin of some of the white blood corpuscles and of the unusual number seen in inflammatory processes. He has carefully studied the formation of the blood clot, which he finds forms gradually, and presents a stratified aspect, having the appearance of a column of red blood surrounded by fibrine and white corpuscles coiled up in the vessel. The tendency of a slowly moving blood column, to retain a shape impressed upon it, can be demonstrated under the microscope in the tongue of the frog. The apex is composed of a homogeneous clot containing some large ovoid cells. The blood clot acts merely as a foreign body, producing a certain amount of circulation, and finally disintegrates slowly. The

white corpuscles and the wandering cells take no part in the organizing process. Zahn [194] has recently performed a series of experiments somewhat similar to that of Shakespeare's fifth series.   A
strong silk ligature was applied to the carotid or femoral of a rabbit,
and, a minute later, removed. Vermilion granules were occasionally applied outside the vessel.   The intima and three quarters of
the media were found cut through.   Endothelium began in a day
or two to form at the edges of the rent, and later grew into it, and
finally filled it up.   There was no cell proliferation in the media,
and the muscular fibres remained unchanged.   The adventitia was
at first hyperemic, and a cell-proliferation took place later. The vermilion granules were found, either outside or inside of cells, and extended as far as the media, but never into the cicatricial tissue.   As
no aneurismal dilatation occurred, he infers that the remaining media
and adventitia were sufficient to prevent that occurrence.   G. Simon
has called attention to ruptures, like those in the arteries of persons
who have been hanged, and he infers that, if such individuals were
restored to life, they would eventually die of aneurismal dilatation,
or thrombosis of the vessel.   This author's experiments on animals
show this view to be groundless.   His former opinion as to the participation of the connective tissue of the media was not borne out,
nor does the adventitia take any part in the repair.   He is sceptical
as to whether a white corpuscle can form genuine connective tissues.

A very interesting series of observations have been made by
Thoma [226] on a formation of connective tissues in the deeper layers
of the intima of arteries, under certain conditions, or what may be
called a compensatory endarteritis, such as is seen in the kidneys in
chronic nephritis.   At birth the change takes place in the Ductus
Botalli and in the umbilical arteries, producing an almost complete
closure of those vessels, and spreads out over nearly the whole tract
concerned during fœtal life in the umbilical circulation.   (Nabel-
blutbahn).   It is particularly marked in the descending aorta.   It
consists of a hyaline connective-tissue at some points, containing
large spindle-shaped cells and branching cells anastomosing with
one another.   This is a physiological example of what occurs in
nephritis, the capillary circulation in the kidney being to a great
extent destroyed; the arteries are too large for their purpose, and
are thus correspondingly diminished in calibre.   In amputation
stumps he finds the vessels narrowing from their point of origin,
finally becoming continuous with a more or less solid cord, which
reaches to the cicatrix; much less frequently the cord is patent in

its entire extent. It may or may not be narrower than the vessel of the opposite limb; in this latter case it ends as a blind cul-de-sac, from which a bunch of small arterial twigs spreads out in the cicatricial tissue.

Verneuil's measurements showed that the vessels regularly narrowed up to their origin from the aorta and vena cava. The three important changes which occur in these vessels are: 1st, a contraction of the media; 2d, compensatory endarteritis; 3d, atrophy of the muscular coat. The same changes occur after ligature in the continuity, but to a less degree, owing to the collateral circulation. His observations confirm those of Schultz on the ultimate shape and size of the cicatrix. The contraction of the vessel is effected, according to Thoma, through the Pacinian bodies which are found in all parts of the arterial system, and lie in the outermost zone of the adventitia; they are able, therefore, to perceive the slightest vibration in the arterial walls and transmit a stimulus to the nerves of the muscle of the vessel. The closure of an artery, whether physiologically or as the result of an operation, depends upon the slowing of the current, but not upon the lateral pressure of the blood. It begins with a contraction of the media. If this be sufficient to restore the normal rapidity of the circulation, further changes are confined to an atrophy of the muscular walls sufficient to correspond with the diminished calibre and diminished tension. If this should not suffice, it is supplemented by a compensatory thickening of the intima.

According to Senn,[232] whose experiments were made upon sheep, the thrombus is accidental, never undergoes organization, and takes no part in the obliteration of the vessel. The final cicatrix is the exclusive product of connective tissue and endothelial proliferation. Permanent obliteration takes place in arteries in from four to seven days. He strongly favors the catgut ligature, which, applied according to his method, does not destroy the continuity of the vessel, and even adds to the strength of the extra vascular cicatrix. In many of his cases the internal tunics were intact. Adopting this precaution, and applying two ligatures, the inner surface can be brought into contact over a larger area and a more extensive surface for cicatrization can thus be utilized. This he regards as a special advantage where the vessel must be tied near a collateral branch.

Wyeth [239] also thinks that division of the inner and middle coats is unnecessary. He has specimens from animals and man showing

3

successful occlusion of the vessel without division of either of the tunics. It is, he considers, a safer method. The thrombus disappears by fatty degeneration, and permanent closure is effected by cells of the intima.

The question of placing a double ligature upon the vessel and dividing the artery between them has recently been revived, and is of interest from its bearing upon the process of repair. Walsham adopted this method, employing as ligature carbolized nerve. He considers the amount of separation of the artery from its sheath a factor as important in influencing the result as tightness of the ligature or division of the coats, since the vitality of the artery depends, in a great measure, upon the blood-supply received from the sheath. For this reason, Savory advocates opening the sheath with the knife instead of the director. The blood supply of the distal side will be more affected than that of the proximal side, since the vessels run in that direction. He cites a case where a fragment of the vessel came away with the catgut ligature, in illustration of this point.

In reply, Holmes finds an objection to this procedure in the free dissection and exposure of the vessels necessary. The interference of the blood-supply from the vasa vasorum is very transitory. " If an artery be securely tied with material which will keep its hold on the vessel until the seat of the ligature is buried in a mass of new fibroid material, secondary hemorrhage, if not impossible, is, at least, very improbable." The method, he says, passed out of use when secondary hemorrhage was common, to be brought back when it is rare. Of interest in this connection is a quotation from Schultz, who says that, if the internal coats are cut, the adventitia has to support the blood column, and is put on the stretch by the retracting vessels. If these are softened by inflammation there may be perforation. Suture of the sheath of the vessel may give an extra support, or the application of two ligatures and division of the artery may allow it to retract. Walsham also points out that if the ligature should cut in this way, the transverse incision assumes a diamond shape, and, if the connection of the internal thrombus with the vessel-wall were slight, it would be disturbed. The method is an old one, and was looked upon with favor by Jones, who says that, " in the single ligature, although the knot is soon covered up and protected by an effusion of lymph, it is placed in the centre of a portion detached from the surrounding cellular membrane, and the process of repair cannot go on so well, since the nutritive vessels are cut off. In the double ligature the knots are placed where the

connection of the vessel with the surrounding tissues is complete."
It had also the sanction of Abernethy, and still has that of Maunder.

Torsion, as has already been shown, was known to the ancients,
and was employed by certain surgeons during the Middle Ages,
but in more modern times, it was not recognized apparently by the
surgical world until brought to the notice of the leading French
surgeons of the early part of the present century.

Velpeau,[83] relates that Professor Grossi of Munich, being at Paris
in 1826, mentioned the fact that his colleague Dr. Koch had not
used ligatures in amputation for twenty years; precisely what the
method employed by him was, is not stated. This conversation
led to experiments upon animals in which Velpeau found that simple
pressure with the tip of the finger at the ends of vessels was sufficient
to prevent further bleeding. He also found abundant confirmation
in literature of the fact that the circulation was not entirely under
control of the heart, and that local conditions often sufficed to enable
even large vessels to be injured without bleeding. Experiments
were also made in "froissement," "fermeture," "renversement,"
and "torsion." He tried the latter method in consequence of his
experience, when a student with a veterinary surgeon, in spaying
cows, the pedicle being twisted after the ovary had been removed.
He also tried it in castration. The method employed by Velpeau
is thus described: "After having seized the vessel by its extremity,
I separate it from the surrounding tissues, and grasp it at its deepest
point in the wound with another forceps to hold it firmly while it is
turned on its axis three to eight times by the first pair of forceps."
Velpeau appears to have employed the method in several amputations. "It is not to be denied," he says, "that the ligature will be
as easy to use as torsion, and perhaps it will be more serviceable in
the hands of the majority of practitioners." He adds "that it may
be preferred in those cases where it is desirable not to leave a foreign body in the wound; animal ligatures would not be less valuable
for this purpose, if we would undertake to try them, than torsion."
He recognizes the disadvantages of torsion in diseased vessels, and
also that small vessels are not easily isolated. According to Bryant,[59] Amussat[107] was the first to communicate a paper on this subject to the Academy in 1829. He was led to employ it from his
experience with torn vessels. Certain it is that the method was
taken up by some of the best surgeons of France and Germany,
but later fell into disuse. In England, little notice was taken of it
in spite of a paper by Costello in 1834. The effect of torsion, Bry-

ant states, is a twisting of the elastic fibres of the adventitia about
the end of the vessel, and a retraction and incurvation of the middle and inner coats.   The twist in the outer coat is permanent, and
cannot be unfolded by any legitimate force.   The middle, and the
inner coats, are retracted in the direction opposed to the blood-
stream, approximated, and overlapped.   They sometimes assume
a nipple-shaped projection; at others, a valvular form, not unlike
the semi-lunar valves of the heart and closing as perfectly.   In some
cases, again, they appear to split; in all, the coagulation of the
blood is favored.   The ampulla-like dilatation of the vessel seen at
the proximal side is attributed to the fact that the coagulation forms
here first, and prevents that close contraction of the parietes which
goes on in other parts, and thus an apparent dilatation is formed.
The safety from hemorrhage in torsion rests upon the twist of the
external, the retraction of the internal coats, and the coagulation
down to the first branch; in acupressure, the permanent safety de-
pending upon the last alone, temporary protection being afforded by
the needle.

Kocher's experiments demonstrate numerous and irregular laceration of the inner coats, over a considerable distance of the wall,
and independent of one another; in ligature the rupture being circular and very close to the ligature.   There is also the narrowing
of the lumen.   Owing to these peculiarities, torsion enjoys, with
acutorsion and acupressure, the great advantage of favoring a rapid
coagulation, and also has its own special advantage of producing
a much more intimate union between the thrombus and the vessel-
walls.   The thrombus plays here, also, the most important part in
hemostasis.   These advantages are specially true of illimited torsion; when properly done, it is the ideal for small arteries.   Torsion
is not probably available for larger vessels, on account of the forces
of the current, which would untwist the vessel; but this objection
does not hold good in acupressure which, in Kocher's opinion, is
likely to supersede the ligature.

Shakespeare [213] confirms, in the main, Bryant's description of the
mechanical effect upon the vessel, but adds that, by pressure of the
limiting forceps, the internal tunic of the artery is rubbed together
a little distance above the end of the arterial stump, producing a
result similar to that already described in what he calls the " modified ligature."   " This is the point where the healing process is
again most active, where the granulations spring from the proliferating intima, and where, by the union of the latter, and the subsequent

changes which have already been mentioned, the lumen of the vessel is first permanently closed."

Ogston [169] found that the calibre of the vessel was closed by the intermingled ends of the fibres of the external coat, forming a thimble-shaped termination enclosing the ragged ends of the middle and internal coats, which were irregularly folded back for the space of a quarter of an inch or so into the bore of the artery.

Sir James Simpson, [179] contrasting the healing of stumps with the rapid union following operations for the relief of vesico-vaginal fistula, in spite of the contact with urine and leucorrhœal discharges, saw, in a ligature, one of the chief obstacles to healing by first intention. It not only acts, he thinks, as a foreign body, but cuts through two of the coats at the time of its application, and eats through the outer coat by the process of ulceration, mortification, and suppuration. In earlier times, attempts were made to overcome this difficulty, by including portions of the surrounding tissues in the ligature, but this was found unnecessarily severe; hence the rule of including nothing but the vessel. Later, large and flattish ligatures were employed. These ulcerated slowly, and were afterwards replaced by those as slender and small as was compatible with due strength. Afterwards endeavors were made to reduce the bulk of foreign body by cutting off the end of the knot. To cut both ends was also proposed, but the procedure was not successful. Animal ligatures were tried, and metallic ligatures also, but Simpson found that they usually excited too high a degree of inflammation. It was for these reasons that he proposed the substitution of acupressure for the old method, basing the use of the needle on what he calls the " great pathological law of the tolerance of living tissue for the contact of metallic bodies imbedded in their substance." Simpson thought that the inner coats were not divided as in the ligature. Bryant denies that the ligature portion sloughs, and is discharged; on the contrary it becomes adherent and vascularized, as surgeons who have been obliged to open an amputated stump must have noticed.

Hewson [158] reports a study of several specimens of human arteries upon which acupressure has been performed. The opposite surfaces of the internal coat were glued together by lymph, the exudation both inside and outside of the vessel being very extensive. There was no clot beyond the point of pressure, and no laceration of the internal coat.

According to Kocher, [166] the first effect of the needle is to produce

longitudinal slits in the intima, but not to the same extent as in torsion. Both in acupressure and acutorsion the walls of the vessels
are thrown into folds and pressed together, and the subsequent inflammation thickens them and glues them together, so that, when
the needle is removed, they still preserve the shape given to them,
and offer a sufficient obstacle to the flow of blood. At first there is
no thrombus, and a fine probe can be passed in through the end of
the vessel, but, as the blood forces its way between the folds, the
rents favor coagulation, and a clot forms at this point, resembling a
cork, which does not project beyond the neck of the bottle; toward
the lumen it is concave, and on the other side convex. This probably occurs earlier than in ligature, which, according to Cooper, takes
sometimes forty-eight hours. In acutorsion the vessel is narrowed
for some distance from its end, and coagulation is still further
favored by this circumstance. Gradually the walls yield, and as
they separate, the thrombus becomes wider and larger, but retains
a firm hold, owing to the slits, through which, eventually, vessels
find their way from the thrombus, which now has become organized.
The ligature could not be removed as early as the needle, as the
action is more local; it cuts through both intima and media, and the
adventitia is so constricted that inflammatory swelling cannot take
place, and sloughing is the result. In acupressure in the continuity,
the proximal and the peripheral ends of the thrombus are continuous, as are also the walls of the vessel, which are thickened by a
connective-tissue growth, a vascularized tissue extending from the
thrombus into the walls.

Shakespeare experimented with the third and fourth methods of
acupressure, and found that the process of healing was very similar
to that which secures obliteration of the artery after ligature. In
the fourth method, or acutorsion, the process was more active.
The needles were allowed to remain in until the specimen was
placed in alcohol, and, in the drawing given (of a specimen thirty-
six hours old), there is no such patency of the end of the vessel as
is described by Kocher, nor is there any such thickening as the wall
at the point of pressure as he represents. Moreover, the thrombus, instead of being concave toward the lumen, as Kocher says, has
its usual ovoid shape.

Ogston has tested, mechanically, the comparative strength of
arteries secured by the methods of ligature, acupressure, and torsion. By experiment, he estimated the internal blood-pressure in
the human subject at from two to eight pounds to the square inch.

Arteries were now taken from a fresh cadaver and subjected to ligature, torsion, and acupressure, and then attached by the cardiac end to a column of mercury. It was found that a column one hundred and fourteen inches in height was insufficient to rupture the ligatured artery. In twisted vessels the artery unfolded at an average height of thirteen inches (or 6.5 pounds to the square inch pressure), *i.e.*, not up to the requisite resistance. He concludes that it would appear likely that vessels secured by torsion are very liable to secondary hemorrhage, especially when the heart, recovering from the immediate shock of an operation, begins to beat more firmly. The acupressure, the fourth method of Pirrie, (which he considers the strongest,) was tried. The column of mercury showed an average of twenty-three and a half inches. It would, therefore, seem a more trustworthy method than torsion, and less than ligature. Torsion may, however, be used in small vessels when the thickness and contractile power of the muscular coat, as well as the comparative amount of tissue included in the forceps, favor its employment. Acupressure and torsion are also considered by Shakespeare as inferior to ligature, owing to the greater slowness of the healing process. Kocher suggests that acutorsion and acupressure would be well adapted to diseased arteries.

LIGATURES.

It is due chiefly to Jones's investigation on the form of the ligature that the modern single thread found, at one time, almost universal adoption. Cutting short both ends was adopted as long ago as 1798, by an American naval surgeon, and by Dr. Maxwell, of Dumfries; and the practice was followed by Hunter and other military surgeons. Hunter also suggested the use of the hair ligature.

The introduction of the animal ligature is generally ascribed to Physick, whose ligatures were made of chamois leather, rolled on a slab to make them hard and sound. Sir Astley Cooper tried them, and they were applied in this country to all the large vessels by Dr. Jamieson of Washington. The latter advised using the buckskin soft, and a little broader than the thickness of the skin. With the introduction of antiseptic surgery animal ligatures have probably largely superseded all other methods of securing vessels. The mechanical action of catgut, kangaroo, or whale tendon upon the vessel-wall differs little from that produced by silk. Barwell has recently proposed, in the treatment of aneurism, the use of a ligature which need not ulcerate through the artery, thus curing aneurism without division of the two internal coats. His flat ox-aorta ligature is intended to act in this way. He thinks it may become organized, having seen remains inseparably mixed with the surrounding tissue fifteen months after it had been applied. In other cases it is absorbed. It is interesting to note that, in specimens where the ligature has been thus employed,—that is, when the intima had not been injured according to Mr. Barwell,—clots have never been found to exist. Dent [220] reports a case of the application of the tendon-ligature to the carotid and subclavian, followed by death ten days after the operation. In an examination of the carotid, the knot of the tendon-ligature was seen in close contact, encysted in a small cavity in the effused lymph. The knot was almost gelatinous in appearance, but small glistening tendinous bands could be seen crossing the dark space. The vessel was completely occluded for a quarter of an inch, being represented by a cord of fibro-cellular tissue. Transverse sections showed that the external coat of the artery "was not ulcerated," and this condition is considered to be due to the slight swelling and softening which such a ligature undergoes. New blood-vessels were found developed in those parts of the tendon which lie close to the artery; that is, rows of spindle-shaped cells, with spaces between, were seen branching in the tendon-tissue, and blood-vessels were seen, passing into the artery and tendon both. The ligature was also infiltrated in other portions with granulation cells. Some of the adjacent muscular coat was being attacked and eaten away.

Arnaud [214] experimented with carbolized catgut on the femoral arteries of dogs; at the end of four days, in one experiment, no trace of the catgut could be discovered; on the other hand, in another experiment, it was seen slightly altered as late as the seventh day.

Lister, [272] in an address before the Clinical Society of London, gives the following example of the dangers attending silk or hempen ligatures. Six hempen ligatures were placed upon the thyroid vessel in an operation for the removal of a goitre; and, although perfect asepsis had been preserved, they all came away at the end of one to eight months, having caused the formation of a slight amount of pus about the knots. The interstices of the thread were loaded with a form of micrococcus occurring in groups of two or three, instead of chains, to which he has given the name granuligera, a form frequent in antiseptic wounds. They produce an acid fermentation, and, in this case, the acid serum became a source of irritation. He uses the catgut ligature, as at present prepared, in a simple reef-knot, tying it sufficiently tight to cause a giving way of the internal and middle coats. He concedes to Mr. Barwell that the rupture of these coats is not essential, but it is advantageous " by leading to a salutary corroborative process of repair." The catgut is prepared from the sub-mucous cellular coat of the intestine of the sheep, the muscular and mucous coats being scraped away. It is then steeped in a one to twenty solution of carbolic acid mixed with a weak solution of chromic acid. As to what becomes of the material, he says: " If it has not been properly prepared, the substance of the catgut becomes converted in the course of a very few days into a soft pultaceous mass, which, when we examine it by a microscope, we see consists of remains of the old cellular tissue of the submucous coats, with the interstices among the fibres filled with cells of new formation. The catgut-tissue is infiltrated with young growing cells, and it is obvious that it is this infiltration which is the cause of the softening; but, on the other hand, if the catgut is properly prepared, instead of being infiltrated by the cells of new formation, it is only superficially eroded. Until nearly a fortnight has elapsed erosion does not begin. It then proceeds gradually, and therefore the thicker the catgut the longer is the time required for its complete renewal. We may fairly consider that from a fortnight to nearly three weeks is long enough for the persistence of a ligature upon an artery in its continuity." The specimen of so-called organization of the ligature upon the carotid of the calf in which " ligatures of new formation are incorporated with the external coat of the artery " is thus explained. The catgut does not come to life again, but, as in the organization of the blood clot, " new tissue forms at the expense of the old, * * * as the old tissue is absorbed by the new, and, * * * as the old is absorbed, new is put down in its place."

Bruns thinks that many failures quoted by him are due to the failure to rupture the inner coats, and the consequent absence of a thrombus. This writer recommends, strongly, a temporary ligature passed through a fine silver tube, and removed in one, two, or three days. He concedes, however, the advantages of the catgut ligature. He corroborates Lister's views as to the formation of a new tissue in the place of the catgut, and sees in this a preservation of the continuity of the vessel, and the absence of the undesirable ulcerating process produced by the silk ligature. The ordinary silk or hempen ligature, if cut short in wounds treated antiseptically, is rarely seen again, and the instances are too frequent to specify where careful search has failed to discover them in the newly formed tissue. When it is necessary to apply the "mediate ligature," as on the pedicle in ovariotomy, its greater holding power gives it a superiority over the catgut, and the testimony of ovariotomists as to its influence upon the healing process is not unfavorable.

Metallic ligatures have not been favorably received, though satisfactory results have been obtained from their use. Holt,[138] for instance, has used a number of wire ligatures in the same wound, the ends being cut short, and the wound healing by first intention; no portion coming away. Pollock[170] used them somewhat like the acupressure needle in the flaps of the wound, twisting the knot on the outside of the skin. They were left in, on an average, five and a half days, with good results. Holmes, on the other hand, objected to them on account of the difficulty of regulating the tension of such a ligature; and Simon, of Heidelberg, also complained that they caused too much injury to the vessel. The tolerance of the tissues for metallic substances left permanently in bones, or in the pillars of the rings, in operations for hernia, has been abundantly testified to.

A word in conclusion upon the action of the blood in coagulating in arteries. According to Conheim, fibrine is formed by the union of two fibrine-generators, fibrinogen and paraglobulin with the cooperation of fibrine-ferment. Fibrinogen is found in the blood plasma, while the fibrine-ferment and the paraglobulin are, for the most part, found in the white blood corpuscles. It is only by the breaking down of the latter that fibrine-ferment and the paraglobulin are set free, and are able to act upon the fibrinogen. So long, therefore, as the corpuscles remain intact, coagulation cannot take place. Physiologically, we have an occasional breaking down of corpuscles, but the living wall seems to possess the power to destroy the small quantities of fibrine-ferment. Endothelium, being common to all vessels of whatever size, is probably the structure which exerts the power shown by the vessel-wall in preventing coagulation, so long as it remains uninjured and performs its physiological functions. If, however, this cell be injured, coagulation will take place. The thrombosis, which occurs after ligature, depends upon the contact of the blood with an injured intima. Virchow thought that a slowing or stoppage of the current was a factor in coagulation, and that the wall could prevent coagulation only when the blood was in motion; but Baumgarten has shown that a double ligature may be placed upon a vessel, and that the blood included will still remain fluid, if sufficient precautions have been taken to prevent injury to the endothelium during the operation. Baumgarten also claimed that thrombosis did not occur in his experiments if strict asepsis were preserved, even though the vessel-wall was uninjured. In the so-called marasmic thrombi it is always possible to demonstrate a defect in the endothelium at the point of coagulation. The slowing of the current favors coagulation only when we have, combined with it, some such defect as this. Zahn has watched the development of the white thrombus in the tongue of a curarized frog. If a crystal of common salt be placed upon a vessel, the observer will notice, presently, an accumulation of white corpuscles at the corresponding point on the inside. A clump may be formed by the addition of more corpuscles, and, if this be washed away by the current, a second one will form in a similar way, if the irritation of the salt continue. At times, the irritation may be so great as to collect a large number of corpuscles, so as to obstruct the vessel, and, in that case, a number of red corpuscles will probably be found imprisoned in the vessel. Usually, we do not find more than a little hill of white corpuscles at the affected spot. This mass subsequently undergoes a change into fine granules, and shrinks a little; while the cell contours begin to disappear. At the end of twenty-four hours the cell outline has disappeared, and we have a grayish, granular, semi-translucent mass in which it is not possible to demonstrate nuclei with reagents. This material is similar to the tough masses of fibrine obtained from blood by beating it with a stick, which affords a striking example of the action of the white corpuscles in coagulation. The white clot is formed, whenever coagulation takes place in a vessel, where the circulation is still going on during the thrombus formation. All thrombi, dependent upon changes in the vessel-walls, would necessarily be white; as would also a thrombus which plugged a wound in the vessel-wall; for, in this case, if ever, we have coagulation with flowing blood. Where red corpuscles are found in considerable numbers in the clot, it is called a " mixed " clot. The great majority of thrombi are either " white " or " mixed." The thrombus may be disposed of by central softening or by organization. The greater portion of the new formation comes from the newly formed vessels developed from the vasa vasorum, or from the adventitia and surrounding connective tissue. It is not probable that the new tissue is formed from the endothelium. The most important change in the new tissue is its contraction, which, since it is unequal at different points, forms an irregular series of spaces through which the blood flows as through a sponge.

A brief summary is given by Baumgarten of the views held by

previous observers on the question of the formation of the thrombus. Virchow considered that it was caused by the slowing or stoppage of the blood current.

Brücke thought the wall of the vessel the important factor in preserving the fluidity of the blood.

Durante held the inflammatory changes in the endothelium responsible for the coagulation of the blood.

Szuman attributed it to injuries of the intima.

Baumgarten found that a single or a double ligature could be applied without the formation of a clot, if strict asepsis were employed. He also observed in the umbilical arteries of two four-weeks'-old children, fluid blood, or blood which had the appearance of being coagulated post mortem. He therefore does not regard the stagnation of the blood as a cause of coagulation, or a rupture or a growth from the intima, for both of these conditions were found in his experiments. By painting croton oil on the piece of vessel included between two ligatures, he was able to produce a clot at this point; but, although the inflammation continued further than this limit, no coagulation took place except between the ligatures. Inflammation of the wall does not therefore suffice, he thinks, to produce coagulation.

His experiments, he says, confirm Virchow's axiom that inflammation and thrombosis do not necessarily go together, and, he adds, they are not discordant with Brücke's theory, for, according to the latter, an inflamed part does not necessarily cease to perform its functions. If, however, the inflammatory condition go far enough to produce a necrosis, the blood will coagulate. But, inasmuch as we frequently find coagulated blood in inflamed vessels in which the anatomical evidence of necrosis is wanting, as in operations not carried on antiseptically, we must assume that coagulation takes place by an interference with the power possessed by the wall to preserve the blood in its integrity, produced either by a total destruction, or by a functional impairment of the tissues of the wall; or, secondly, by a passage through the wall of chemical substances, which, without altering the structure, may set in operation certain processes, like ferment-formations, which lead to coagulation. The rôle played by the so-called granule masses, granular débris, blutplättchen or blood plaques, in the process of coagulation is at present exciting much attention. The reader is referred to Osler's [235] recent lectures upon this subject.

MINUTE ANATOMY OF THE ARTERIES.

According to Shakespeare,[213] in the larger arteries in the adventitia there is a net-work of branched corpuscles which lies in lymph spaces formed by a loose reticulum and felt-work of white fibrous tissue. Between it and the media, a few elastic fibres, collected into a net-work, are found in the larger vessels of this class. A small number of elastic fibres constitute the elastic layer of the intima; and between this layer and the endothelium are a small number of branched connective tissue-cells, connected together into a membranous network. The media consists of a continuous muscular membrane composed of a single layer of muscular cells.

In larger vessels there is a more complete development of the connective tissue felt-work, whose fibres become more longitudinal; scattered among these bundles, a few fine elastic fibres are found, and also connective tissue corpuscles and lymphoid cells. These loose meshes are lymph-spaces. The elastic fibres become more abundant, and larger near the media, and they here form a more dense net-work, separating the adventitia from the media, and known as the external elastic membranes. The media has a large number of muscular fibres, the cells being arranged in transverse bundles. The different layers are separated by plates of elastic tissue in the form of fenestrated membranes, running mainly longitudinally, connected together by net-works of fine elastic fibres. A small amount of fibrous connective tissue is formed between the bundles. At the external boundary of the intima is another dense collection of elastic tissue, the internal elastic membranes consisting of two or more fenestrated elastic layers, so closely packed against each other as to present, in section, an appearance of a structureless membrane. (According to Winiwarter, the normal intima of some vessels of this size, as the tibial artery, possesses two well-formed laminæ with an intervening fibrous layer in which spindle and stellate cells are found.) On the internal face of the elastic membrane is a slight accumulation of white fibrous tissue arranged longitudinally; the fibres intercross, however, leaving lymph spaces between them, containing fusiform and branched connective tissue corpuscles. This layer is covered by the endothelium composed of lozenge-shaped plates. In transverse sections the elastic membrane has a wavy outline.

In the larger arteries the number of muscular layers is increased, as well as the thickness and size of the elastic plates. We find, in some vessels, longitudinal and oblique bundles of muscle, especially in the inner portion of the media. They are also found in the adventitia, but rarely in the longitudinal fibrous layer of the intima.

(Thoma has observed muscular fibres in the deeper layer of the intima of the aorta, at the neighborhood of the origin of the great vessels of the head and upper extremities; this layer bears a direct relation to the branch. Whether these cells should be regarded as part of the intima or media is a question; they are, however, inside the elastic lamina which, at other points, is the line of demarcation between the intima and media. These fibres are sometimes longitudinal, and sometimes transverse, in direction. A striking example of this arrangement is seen at the bifurcation of the aorta, where bundles of longitudinal fibres are found partly in the intima, and partly in the media. The evident intention of this band of fibres is to support the spur, formed at the point of division, against the blood-stream. In a less pronounced form, they are to be found at all arterial divisions. He finds the media in the aorta of dogs exceedingly rich in muscular fibre.)

The elastic lamina of the larger arteries is much thicker, and is sometimes laminated; so that we see, apparently, two laminæ with a small amount of connective tissue between them. The fibrous layer of the intima is now quite distinct.

The muscular cells of the arteries are, in the main, simple, smooth, fusiform cells with rod-shaped nuclei. In the larger trunks the ends may be more or less bifurcated, or even branched. In the aorta, flattened, stellate, muscular cells are often met with.

In the large vessels, the outer and middle coats are supplied with blood-vessels, the vasa vasorum. In a few instances, capillary vessels even enter the tunica intima.

CHAPTER II.

A CRITICAL study of the investigations on the nature of repair in arteries after ligature, leaves one strongly impressed with the fact, that those which have been made since the beginning of the histological era have been conducted chiefly with reference to their bearing upon prevailing theories of the day, and from a pathological, rather than from a surgical standpoint. The history of this research is, in fact, a history of the rise and gradual development of the science of cellular pathology.

Since Hunter first enunciated the theory of the organization of the thrombus, this field has been a favorite one on which to test the theory of cell action, advanced successively by Schwann, Virchow, Recklinghausen, Conheim, His, and others. Investigation has been directed to the action of the white corpuscles, the wandering cell, the connective tissue corpuscle, and the endothelium; the early changes which occur in a vessel after ligature being supposed to offer special facilities for such studies. During the last half century, surgeons have been content to leave the question of the repair of blood-vessels mainly in the hands of the histologists, and the more purely surgical aspects of the question have remained pretty much in the condition in which they were left by Jones. It is for the purpose of reviewing the whole question from a surgical point of view that the following series of investigations has been conducted. The attempt is here made to trace the process of repair from the earliest visible change to the development of a tissue which is incapable of further elaboration; in other words, to the permanent cicatrix. For this purpose a series of experiments has been performed upon dogs, covering the most important points, and ranging in time from a few hours to four months. The experiments on horses were made chiefly with a view to obtain specimens suitable for macroscopic study. These investigations have paved the way for a more intelligent study of a series of specimens taken from the human subject, embracing ligatures in amputations, as well as those made in continuity. Finally, an attempt has been made to explain the differences found in the two results of the last series, by a comparison with the changes observed in the obliteration of the hypogastric artery and the ductus arteriosus after birth.

In many of the experiments about to be described, no special attempt was made to keep the wound in an aseptic condition. In the case of ligature of the carotid artery, a compress wet with carbolized water was tied loosely around the neck beneath the collar; or, if the wound was in the femoral region, no dressing at all was applied. Cotton sewing-thread was used for the ligature, and for the sutures.

As a rule, these wounds healed rapidly in the deeper portions, but a small subcutaneous cavity was almost invariably found containing serum, muco-purulent fluid, or pus, as the case might be. Occasionally, a minute fistulous tract communicated from this sac with the ligature, but more frequently the thread was completely encapsuled in the surrounding callus. The wound through the skin had generally healed by first intention. In a few cases considerable suppuration took place, and in one instance secondary hemorrhage followed.

DOGS.

FEMORAL. TWO DAYS.

A double ligature was placed upon the femoral artery of a dog, and carmine granules were freely powdered over the wound. The animal was killed in forty-eight hours, and the specimen, after being hardened, was divided longitudinally into halves. The ligatures were placed too near together to leave a space of sufficient size for the purposes intended in the original experiment, but the distal portion illustrates very beautifully the conditions ordinarily produced by the ligature at this period. (Fig. 1.)

The dense, tendon-like condition of the adventitia and peri-adventitial tissues enclosed in the ligature is well shown. A mass of cells containing carmine granules is found near the ligature, but no granules are found within the vessel. At the same time a number of granulation cells have penetrated the almost uninjured wall of the vessel, and have gained access to the interior. None of them, however, contain carmine granules, which are found in the external granulation tissue. The thrombus, which is slightly laminated, contains a considerable number of white cells. No apparent change has taken place in the cells of the intima. The ligature is completely covered in by granulation tissue.

Remarks.—This specimen shows the relations of the parts soon after the application of the ligature. The protective influence against hemorrhage of the dense external wall is well shown. It is also an illustration of the much disputed statement that wandering cells can find their way through the walls of a vessel. It may be said that, in this case, the walls are not uninjured, but it is quite evident that we have, not a growth of granulation tissue through a cleft in the wall, but a fair example of what wandering cells can accomplish.

FEMORAL. EIGHTY HOURS.

The femoral artery was tied with a cotton ligature, both ends being cut short. The vein was adherent to the vessel opposite the point of callus-formation. The specimen was divided into halves; longitudinal, and cross sections being taken.

Both thrombi are well developed, but the usual difference in size is not so apparent. The media is not curled in, but has been cut completely through. The ligature, still in its proper place, holds the fibres of the adventitia; there are several lateral ruptures of the media near the ligature, to which points the thrombi closely adhere; elsewhere they have separated from the wall. At these points of attachment the staining shows the growth of new tissue into the thrombi, and the amount for this period is considerable. A study of these points with high power shows the lamina ruptured and reflected back, and a cell-growth coming from the inner third of the media, the outer portions of which are normal. At other places, very near the ligature, where the thrombus does not appear to be attached, the lamina is lifted up, pustule-like, by a mass of young cells lying between it and the media; and there are other similar cells in the media also, but they are very few in number. The adventitia is infiltrated with round cells. The intima, at some distance away from the ligature, is normal; near that point there is proliferation of its cells, forming two or three layers of spindle-cells, mingled with round cells, which, when in contact with the clot, project into it, at times, at right angles to the wall. An examination of the thrombus, by picking it apart with needles, shows a spindle-cell and anastomosing stellate cell reticulum in certain portions.

Remarks.—This specimen shows that, where the rupture in the elastic lamina has occurred, the clot is most adherent and the cell-growth most active. No real union, however, has yet taken place; the cell-growth simply holds and strengthens the clot, replacing

4

some of its old substances with young cells. New cells do not come solely from the intima, but are noticeably more numerous where that layer has been broken through and the media exposed.

CAROTID. FOUR DAYS.

The left carotid of a large dog was tied with a silk ligature, both ends of which were cut short. The animal was killed at the end of four days. A longitudinal section showed a large proximal thrombus, the distal thrombus, if any existed, having disappeared.

The walls of the proximal portion are much distended by the clot, and consequently are thinner than normal, and much compressed. The rupture has extended through three quarters of the media, the ends of which are frayed and turned in toward the clot. Connected with this wounded surface, which constitutes a small gaping wound at the fundus of the cul-de-sac (Fig. 2) formed by the proximal end of the vessel, we find a ramifying net-work of fibres and round cells, which extends for some distance into the thrombus, but is most abundant near the wounded surface; it comes in contact with the inner wall at one or two other uninjured points, however. It takes staining readily, and has the appearance of young growing tissue. There is no infiltration of the media with wandering cells. No change is to be detected in the endothelial layer. The ligature is unaltered, and includes the fibres of the adventitia in a dense, transparent, almost homogeneous bundle. It is surrounded by a growth of small round cells which occupy the periadventitial tissue for some distance above and below the point of ligature. Cross sections show that, near the ligature, the cells have penetrated into the inner layers of the media.

In the distal portion, the media is ruptured and thrown into longitudinal folds. There is no infiltration of the walls with round cells. The most noticeable point is the endothelium, which is in a state of proliferation. (Fig. 3.) A high power shows a number of spindle-cells disposed in no regular order, hanging generally by one extremity to the wall, or interlacing one another, and also stellate cells, and mother cells containing numerous nuclei.

Remarks.—The apparent new growth of young tissue to be found in the proximal thrombus consists of cells resembling white corpuscles and fibres of coagulated fibrine. The cells are probably, chiefly the white corpuscles of the clot, which accumulated during the process of coagulation. It seems probable that a few cells may have wandered into the clot from the exposed deeper layer of the media;

and possibly, a few came from the adventitia. The endothelium is firmly compressed by the thrombus, and gives no evidence of any change. In the distal portion, however, where no thrombus was seen, it having probably been lost in the preparation, a distinct proliferation of the cells of the intima is observed.

CAROTID. ONE WEEK.

A cotton thread ligature was passed around the right carotid, and cut short. There was a slight collection of sero-purulent material beneath the wound in the skin which had united by first intention. The ends of the vessel were imbedded in a well formed callus, in which only a few traces of the ligature could be found. The thrombi were about one half an inch in length, although the proximal thrombus was a little the shorter of the two, but, as usual, considerably broader. They were slightly laminated in the portions near the apex of the clots.

Proximal End.—The thrombus is firmly attached at its base, and loosely at its apex. It is, at points, pervaded with anastomosing masses of protoplasm with ill-defined structure. At others, a round cell-accumulation is seen at the sides of the clot. Near its apex, the clot, which has here shrunk away from the wall of the vessel, is still attached to columnar anastomosing masses of young cells. This growth consists of round and spindle-shaped cells. Although they apparently originate from the intima, an inspection with high power shows rents in the lamina (Fig. 6) through which some of the cells have evidently grown. The other portions of the intima which are visible, show a slight thickening of its layer of cells, which are round and spindle-shaped. At the point of ligature, the walls of the vessel are slightly retracted, and granulations are beginning to grow into the thrombus.

Distal End.—Here, also, the walls have retracted, and the thrombus is invaded by granulations. The thrombus is narrow, and loosely attached to the sides of the contracted vessel; and, where it has separated, an increase in the cells of the endothelial layer may be seen. At one or two points, ornamental festoons of spindle-shaped cells have been frayed out from the wall by the contracting clot. At the apex of the thrombus, there is an accumulation of round cells. Near the ligature, the clot is, in places, separated from the wall by fluid blood.

Remarks.—We have here a somewhat earlier invasion than usual of the lumen of the vessel by granulation tissue. The growth of cells from the inner walls of the vessel has, in places, permeated the clot for some little distance, and serves to hold it firmly in place. This is particularly noticeable near the apices of the thrombi. Much of the cell-growth, which appears to spring from the intima, is found to project, through ruptures in the elastica, from the media.

A series of arteries were tied with special reference to antiseptic precautions, and their influence upon the formation of the thrombus. All details of the antiseptic treatment were rigidly adhered to. A powerful double-nozzle carbolic steam-spray was used. The hair was shaved, and the skin washed with one-to-forty carbolic wash. All instruments were placed in carbolized water. Chromicized catgut was used chiefly as ligatures; but cotton thread was occasionally substituted. The dressing consisted of an application of iodoform powder, over which was placed a thick layer of borated cotton, which was held in place by a bandage soaked previously in carbolized water. The dressing was not changed, but iodoform was powdered on freely around the edges daily. No difficulty was found in keeping a dressing over the carotid or femoral arteries for a whole week, without change.

The following cases are selected from a large number, for the purpose of illustrating this point:

CAROTID. ONE WEEK.

The common carotid was tied at about the middle of its trunk with catgut. On opening the vessel above and below this point, no thrombus was at first seen, but, after laying open each end of the vessel to the exact point of ligature, two very minute thrombi were observed which might easily have been overlooked. The callus was not so large as in non-antiseptic cases. In a second case, the catgut was placed very gently around the artery so as to close the lumen, if possible, without doing injury to the walls. Complete union by first intention took place in the wound. On dissecting out the vessel, it was apparent that the callus was exceedingly small, being barely sufficient to cover in the ligature. On opening the vessel carefully with fine microscope-scissors no thrombus was found; but, on exploring with the finest bristle, it was discovered that the lumen had not been obliterated, and the blood still flowed through the ligatured point. Cross sections of this portion of the vessel were made for microscopic study, and disclosed the fact that, although the walls of the vessel had not been accurately approxi-

mated, a number of ruptures through the elastic lamina had taken
place, and, from these points, granulations were sprouting into the
lumen of the vessel. In still another case, a cotton thread was tied
around the femoral artery. The wound healed throughout by first
intention. The callus was of about the size observed in cases oper-
ated upon without special antiseptic precautions. To make sure
that the vessel-lumen was completely obliterated by the ligature,
water was injected into one end with a fine syringe, but it failed to
pass through. On laying open the vessel a minute thrombus was
found at each end, the proximal being slightly the larger. Each was
about the size of mustard seed.*

CAROTID AND FEMORAL. NINE DAYS.

The carotid and femoral of a medium-sized dog were tied with
cotton ligatures without special antiseptic precautions: the ends
were cut short, and the animal was killed on the ninth day. The
ends of the carotid were separated from one another about one
third of an inch, the knot still being attached to the distal portion;
the two ends being enclosed in a mass of inflammatory tissue. A
longitudinal section divided the specimen into halves.

The chief point of interest is the shape of the proximal thrombus,
which is about a quarter of an inch in length, and resembles a polyp
attached by a slender stem to the end of the stump of the vessel
inside. It has a twisted appearance, as if in a more plastic state, it
had been twirled about on its own axis by the motion of the blood.
Surrounding the stem is a clear homogeneous blood-clot, in which
cell-growth can be readily studied. At the bottom of its cup-shaped
cavity, which is lined with frayed and bruised portions of the media,
there is a slight growth of delicate anastomosing stellate and spindle-
cells with prolongations, making a reticulum of web-like structure, in
the meshes of which red corpuscles are lying. This tissue is attached
to adjacent fibres of the media, which is considerably altered and infiltrated with inflammatory cells. A little higher up the cell-growth
is more voluminous, and the delicate mesh-work is obscured by
masses of young cells. Occasionally columnar shaped masses of cells

---

* Circumference of vessel inside above proximal thrombus. 3.40 mm.
Diameter of vessel therefore, about . . . . . 1.08 mm.
Proximal Thrombus, length . . . . . . 1.61 mm.
 " " breadth . . . . . 0.65 mm.
Distal Thrombus, length . . . . . . 1.49 mm.
 " " width . . . . . . 0.56 mm.

are seen extending some distance into the transparent clot. A little higher up, the intima is found slightly thickened, and a few spindle and stellate cells are seen projecting into the clot from this layer. The thrombus is attached by its stem to the delicate anastomosing cell net-work just mentioned, and elsewhere unattached, except at one point, where it appears to lean against the wall; it is composed of red corpuscles and fibres of fibrine and also of a cell-growth, which readily takes the coloring matter used: this new tissue, if such it be, is arranged in columnar-shaped masses.

The distal portion is still surrounded by the adventitia which is held by the ligature, but, at one point, there is a slight rent through which some granulation-cells are pushing their way. The lumen is occupied by a transparent homogeneous clot, into which we find, near the end of the vessel, a moderate number of round spindle-cells growing; and, at points where the media has been lacerated, columnar or granulation-like masses of cells are seen. The intima shows here and there, a slight activity, and some of the cells are evidently produced by this layer; but, as a whole, the cell-growth is slight in amount in the distal end.

Remarks.—This specimen is of interest on account of the transparency of the clot and the period of organization of the cicatrix. The granulation tissue has not yet penetrated the vessel to any great extent, although the walls of the proximal end are infiltrated with granulation-cells. The delicate cell growth in the interior is limited in amount. The columnar masses of cells appear to originate from the granulation tissue.

The thread at the time of ligature had cut its way so nearly through the femoral artery, that the walls had been severed, and a clot had formed between the ends.

The walls of the proximal ends have contracted, and an abundance of granulation tissue has grown into the thrombus, which is a large one. The granulations have taken the columnar form. In the peripheral end the clot is not present, and there exists a growth of a pyramidal shape composed of columnar masses of cells intertwining in such a way as to produce spongy, or cavernous, tissue. (Fig. 8.) This tissue appears to spring from a cleft in the coats of the vessel at the point of ligature. It is composed largely of round cells, but both the spindle and stellate forms are found.

There is a slight proliferation of the endothelium, spindle and round cells being found in this layer. The walls of the vessel, in the immediate neighborhood of the thread, are much infiltrated with round cells.

Remarks.—We have here an example of the tissue which displaces the blood clot, and an indication of the way in which it acquires its peculiar cavernous character: we see also its origin from tissue which has penetrated the open end of the vessel.

FEMORAL. TEN DAYS.

A single cotton thread was applied to the femoral artery, and cut short. A well-marked callus was found. On laying open the specimen by a longitudinal incision, the ends of the vessel were found retracted some distance from each other, the ligature remaining in contact with the distal end.

The proximal end has apparently cicatrized, the ends of the media being in close apposition, and apparently forming a perfect cul-de-sac. Longitudinal sections show the thrombus lightly attached by a pedicle to the fundus of this sac. (Fig. 9.) It has a polypoid shape, and is surrounded by an unstratified fresh clot. The intima is not obscured by thrombus, and can be conveniently studied. At the line of union of the two sides of the media the cells of the endothelium have proliferated, and formed a thickened layer. (Fig. 7.) At some points, where there is a slight, excavated wound the cells had grown down, and across, the depression. (Fig. 10.)

The media, on close inspection, is found infiltrated with wandering cells on either side of the point of ligature, but these cells have not yet reached the interior. (Fig. 9.)

The distal portion has retracted considerably and its walls have separated, giving an opportunity for the granulation tissue, which surrounds the still unabsorbed ligature, to grow freely into its lumen.

Remarks.—The appearance of the proximal portion in this specimen is instructive as showing the condition of supposed union by first intention, but a critical examination shows that only the first stage of the healing process has been completed, and that the infiltration and absorption of the ends of the media (now apparently united) are just beginning. It also shows that the endothelium is capable of proliferation and of taking part in the healing process, but that the amount of tissue produced by this layer is comparatively insignificant.

In a second specimen of ten days' repair, the artery, the carotid, had a large proximal thrombus with a small extravasation at the site of the ligature. The vessel presented the appearance of the so-called "ampulla-like dilatation." The coats of the artery are placed upon the stretch, and the media is correspondingly narrowed. (Fig. 2.)

There is no sign of a proliferation of the cells of the intima, but granulation-cells are penetrating the thrombus at the point of ligature. The distal thrombus is thin, and has shrunk from the walls of the vessel, where a considerable activity of the cells of the intima can be seen. There is also some penetration of granulation-cells at the point of ligature.

Remarks.—This specimen presents more accurately the aveıage condition of the process of repair at this period. The walls are beginning to retract, and the granulation tissue to invade the thrombi. It is worthy of note that, when the clot is firmly wedged in, as in the case of the proximal clot, there is no growth of cells of the intima; but when the clot is less intimately in contact with the vessel-wall, more or less activity is to be noticed in the cells of this layer.

A critical study of these specimens shows that, up to this period, the walls of the vessels have been kept in contact at each end, but that the two ends have begun to separate from one another, the ligature either having been disintegrated, or cast off, or left attached to one of the ends.

At or later than this time, the ends of the vessel begin to open by a retraction of the vessels, now freed from the ligature, and granulations begin to grow into the thrombus at these points.

In some cases, at points favorable for observation a slight proliferation of the cells of the intima appears to have taken place, but the amount of this growth is small, and difficult to find. In several instances a distinct immigration of wandering cells was observed, establishing beyond a doubt that this much disputed action of these cells does occur, but the total amount of cells collected in this way, in the interior of the vessel, is small. It is evidently the beginning of a process, which eventually ends in an absorption of the edges of the vessel-walls and brings about their retraction, thus giving ingress to larger masses of granulation cells.

The adventitia, which is not ruptured by the ligature, is collected into a bundle of dense fibres, and forms a tendinous band, holding the two ends together: this band is eventually permeated and disintegrated by the granulation cells, allowing the two ends to retract from one another. Granulation tissue has accumulated in considerable quantity around the ligature, and varies in quantity with the amount of inflammation. It forms a callus in which the two ends of the vessel are imbedded like the fragments of long bones after fracture. Even in cases where the strictest antiseptic precautions have reduced the inflammatory reaction to a minimum, this callus-like mass of tissue is found.

When the inflammation has been great, it may assume considerable proportions; but in cases of suppuration and hemorrhage it will, on the other hand, be found to have broken down and to have left the artery more or less exposed.

The thrombus varies greatly in size. In the antiseptic cases it has been reduced to an exceedingly small mass, so small as to be easily overlooked; but in no case has it been absent entirely. Like the external callus it appears to indicate the amount of traumatism which has been produced, but not, however, with the same accuracy, for, in some of the most perfect examples of asepsis, it was larger than in cases in which this treatment had been less successfully carried out. A series of experiments with the double ligature, performed to test the question of the activity of the endothelium and wandering cells, is inserted here, to preserve as nearly as possible a chronological order.

DOUBLE LIGATURE.

ONE WEEK.

The femorals of a large dog were exposed at two points, each about one inch apart. The sheath was opened only at the point where the ligature was applied: carmine granules were then freely sprinkled on the wound. No attempt at an antiseptic dressing was made. The intermediate portion was compressed so as to empty it of blood, but in one case the attempt was only partially successful. There was a suppurating wound beneath the sutures, at the end of a week, when the animal was destroyed, and in bringing the body into the laboratory on that day, a hemorrhage took place from the left femoral at its central portion. A part of the vessel which included the ligatures was removed, cut in halves, and each half split longitudinally.

At the proximal end, from which the bleeding took place, a fresh clot is seen occupying its mouth. This clot is concave on its inner surface which is covered by white corpuscles. The anterior lip of the vessel is everted. The ligature was found in the secretions of the wound. There is no clot in the middle portion, the ends of which are slightly separated, admitting the entrance of granulation tissue for a short distance. There is a purulent infiltration of the media, which has separated it partially from the adventitia, but only a very few cells have found their way through the coats of the vessel into the interior. There is no trace of endothelium. One or two granules of carmine are seen in the granulations at one end, but most of them have accumulated in the upper layers of the granulation tissue forming the wound. The right femoral has considerable clot in the middle portion. The media is infiltrated in its outer half with cells, and some of these appear to have penetrated through the walls. There is an appreciable growth of endothelium: at one or two points collections of round cells are seen, which might have been the result of proliferation of the cells of the intima. Each end of this portion is slightly opened, and occupied by a mass of granulation tissue. The carmine granules are, as in the other case, upon the surface of the wound.

Remarks:—In both of these vessels the ends of the isolated portions were opened somewhat prematurely by the suppurative process; and the chief supply of cells found in this portion came from the granulations which forced their way in at the end, and not from cells wandering through the coats of the vessel (although a vigorous attempt was being made to penetrate them), nor from the endothelium. In one case no change in that layer was found, and in the other a slight proliferation at one or two points. Care had been taken to preserve the vitality of the endothelium by leaving the walls of the vessels as much as possible in connection with the sheath.

DOUBLE LIGATURE.

ONE WEEK.

The following experiment on the femoral artery is recorded, as the operation was performed under the strictest antiseptic precautions. Care was also taken to avoid injury to the walls of the middle portion, and not to interfere with their vascular connection with the vasa vasorum by a separation of this portion from its sheath. The proximal ligature was applied first, and the vessel gently compressed while the distal ligature was tied, so that as little blood as possible should remain between the two ligatures. The wound healed by first intention. On opening the specimen a small thrombus was found between the two ligatures (No. one chromicized gut), and two small thrombi about one sixteenth of an inch in length were found at the distal and proximal ends respectively; the latter being slightly the larger of the two. A moderate callus surrounded each ligatured point.

Middle Portion.—The thrombus is small, and the lumen partially collapsed. Along the inner walls, at intervals, there is a formation of spindle-shaped cells, slight in thickness, hardly more than a single layer. At other points, there is an endothelium of cube-shaped cells, such as are described by Baumgarten. There are a number of ruptures in the lamina elastica, and cells appear to be growing through these spots from the media (somewhat similar to the growth in Fig. 6). The ends of this portion are surrounded with granulation tissue, and the walls at this point, though still loosely in contact, are infiltrated with round cells, which, at one point, have just begun to push their way into the lumen.

Distal End.—The walls have slightly retracted here, owing to a disintegration and yielding of the catgut, and granulation tissue presents at the opening, at one side of which the thrombus remains attached. No change is observed in the intima. The ends of the media are slightly infiltrated with round cells.

Proximal End.—The thrombus is polypoid in shape, and attached to a fragment of the vessel-wall (as in Fig. 9). No endothelial growth is seen, and only a slight infiltration of the ends of the media. Granulation tissue has grown in between the proximal end and the middle portion, and the fibres of the catgut are thoroughly infiltrated with wandering cells.

Remarks.—In this experiment no wandering of cells is observed through the walls of the isolated portion of the vessel. There is a slight proliferation of the endothelium, and a growth of cells from the media through ruptures of the inner wall. Granulation cells are apparently working their way through the ends of this portion of the vessel, and at one spot have penetrated to the interior.

DOUBLE LIGATURE.

FOURTEEN DAYS.

The left femoral of a large dog was tied with two cotton ligatures one third of an inch apart; and the animal was killed at the end of fourteen days. The external wound had healed well, but a wound was found below the surface containing pus, and the remains of the ligatures. Underneath this minute abscess lay the granulation tissue which enclosed the vessel. A longitudinal section shows the proximal thrombus shorter than usual owing to the presence of a branch a short distance above the point of ligature. The central portion is collapsed and contains traces only of clot; the distal portion has a well-formed thrombus; no traces of the threads exist at the ligatured points.

Proximal Portion.—The walls of the vessel have separated and very typical granulations are growing into the deeper layers of the clot, which they have, in a measure, replaced. The cells, of which this tissue is composed, are round and spindle-shaped with a hyaline intercellular substance, and the borders of the granulations are lined with spindle-cells. The granulations, being crowded close together and, overlapping one another, form a papillary growth, and between them enclose, consequently, an anastomosing series of spaces, which communicate with the open cavity of the vessel.

A careful search fails to find a communication of these spaces with the capillaries found in abundance in the external granulation tissue. Above the level of these granulations (Fig. 11 g.), and at a point abreast of the still preserved blood-clot, which is loosely attached at one or two points only to the wall, there is a decided thickening of the intima. On the surface of this thickened layer may be seen normal endothelial cells, some of which seem to be in a state of division; just below them is a layer of spindle-cells, and beneath, lying upon the elastic lamina, are to be found some round and irregular shaped cells, with hyaline intercellular substance and an occasional vascular space. The tissue of this layer corresponds closely with that seen in the granulations with which it becomes continuous. Such an appearance as this is usually explained by assuming a growth of the cells of the intima. Little or no change is seen in the endothelium, and the connective tissue growth, beneath, is directly continuous with the tissue, which has grown into the vessel from without. It is possible that the layer of spindle-cells may, however, be a product of the deeper layers of the intima. This thickening in the intima is the beginning of a growth of tissue, which, in more advanced specimens, is seen to give the peculiar crescent-shape to the final cicatrix. The remains of the thrombus are adherent to the granulation tissue, and to the sides of the vessel.

Middle Section.—The double ligature was applied, the blood having been excluded, for the purpose of determining whether a growth of endothelium could be observed. This segment was found collapsed in longitudinal folds. A very slight amount of clot was seen in that portion.

The ends of the vessel have slightly opened, allowing a growth of granulation tissue to make its way in for a short distance at either end. A careful exploration fails to discover any growth of endothelium. A few large endothelial cells are scattered about in the space, as if desquamated from the inner wall, but no other traces of intima cells are seen. The granulation tissue is rich in capillary vessels which lie in a mass of spindle-shaped cells, and, as they can be traced to the edge of the growing mass, their earliest development can be studied. (Plate IV, Fig. 11-v, Figs. 12 and 13.) There is nothing to indicate a formation of vascular loops or a following out of an anastomosing system of cells. The vessels can be followed until the walls are barely far enough apart to admit a single red corpuscle (Figs. 12 and 13), and are then lost in the surrounding mass of spindle-cells. The mode of formation appears to consist in the disposition of longitudinal bands of spindle-cells, developed from round cells, offering a channel to the blood, which has been liberated from the original vessels by a softening of their walls, the blood forcing its way into the new soft tissue which then disposes itself, in the shape of walls, around the new spaces channeled out. In other words, the new formation of blood-vessels is produced by an intercellar, rather than by an intracellular development. No change is noticed in the media of this portion; the adventitia is infiltrated by the surrounding inflammatory tissue.

The Distal Portion.—The walls of the vessel have slightly expanded to admit granulations, to which the ·thrombus, a small one, is firmly attached. The lower portion of the thrombus (that nearest the lumen) is infiltrated, in certain portions, with new cell-growth, as shown by the staining of those portions. It has shrunk away from the sides of the vessel, where endothelial cells may be observed adhering to the edge of the thrombus; and also spindle-cells, lying in confused masses, they having been, apparently, torn away in the shrinking process, from the walls. No vessels are to be found entering the thrombus, either here or in any other portion of the vessel, through the sides.

Around the three portions we find a mass of new tissue, which extends, upwards and downwards respectively, upon the outer walls of the proximal and distal portion, and completely encircles the middle segment.

Remarks.—This specimen was prepared to study the growth of the endothelium when unobscured by blood-clot. As no special pains was taken to preserve the attachment of the vessel to its sheath, it may be argued that no growth of endothelium was possible, the blood supply from the vasa vasorum having been removed; but the segment was being rapidly obliterated by the granulation tissue from without, showing that it is quite possible for such a result to be arrived at without the aid of the endothelial cells. There is no cell-formation in any of the portions which might not be referred back to the granulation tissue, that is, to sources external to the vessel; but it is probable that some proliferation of the endothelium in the proximal and distal portions had, nevertheless, taken place.

The external and internal callus development is well illustrated in this specimen; and the mode by which the central portion is gradually being destroyed is also indicated.

The formation of capillaries is conveniently studied here also, as the surrounding conditions have necessitated a growth in one direction only. It illustrates also, very perfectly, the mode of "vascularization of the thrombus."

DOUBLE LIGATURE.

ONE MONTH.

Two ligatures were placed upon the femoral artery about one inch apart; before removing the specimen, the limb was injected with Prussian blue. The proximal and distal ends of the vessel were united by a long cord, in the axis of which were found the remains of the central portion. The proximal ligature was found encapsuled, and but slightly altered. The distal ligature had disappeared.

The thrombus of the proximal portion may be divided into two parts for the purposes of description. The upper portion is a typical, stratified clot, of an oval shape, and rests somewhat loosely upon the other portion, which is rapidly being disintegrated by granulations.

The new tissue formed within the walls of the vessel is better studied in specimens taken from the proximal portion (Plate VI.). Here we see that the walls of the vessel have separated, and retracted somewhat; and a mass of granulation tissue has grown into the interior of the vessel. Up to a certain point, which is readily recognized in the drawing, the tissue is precisely similar to that found outside the vessels. It is typical granulation tissue, consisting of cells of many shapes, among which the round, epithelioid, and spindle forms are seen. There are numerous small vessels which take their origin from the vessels of the external growth (Fig. 16 g.). Along this line of granulation tissue, traces of the thrombus are seen, which has already been infiltrated and disintegrated by a growth of new tissue. At the free surface of the lumen, this tissue shows itself in the form of typical granulations, which, when studied with a high power, show the structure delineated in Fig. 18. It consists of a transparent more or less homogeneous ground-work, in which spindle, round, or stellate cells are to be found. The granulations are exceedingly tortuous, and leave spaces in which fresh blood flows. They are covered with a single layer of endothelium. Clumps of pigment may be seen at different points (d). The new tissue extends farther up the vessel on its sides than in the centre: here we find spindle-shaped cells predominating, which are imbedded in a similar intercellular substance; here the endothelium is somewhat thicker, and is in strong contrast to the subjacent spindle-cell layer. The growth extends much farther up one side of the vessel than the other. The internal callus is " eccentric," and this eccentricity is due to the presence of a collateral branch on the side where the growth is shorter.

The injection mass has forced its way freely in between the granulations on the surface, and also into the capillaries of the granulation tissue which enter from below. We have, in the interior of the vessel, two independent channels for the blood to run in, but although a successful injection has been made, it fails to show that these two systems communicate with each other.

Central Portion.—This part of the vessel is evidently in a process of rapid disintegration. The ends have shrunk, and have already been acted upon by the granulation tissue, which penetrates their open mouths. Cross section shows the tube collapsed, and its walls only slightly separated by fragments of clot and granulation cells; at some points we find these cells infiltrating the walls from without; at others, from within. Near the ends of the fragment, there is an abundant growth of young vessels, accompanying the ingrowing tissue. No growth from the intima, as a layer, is seen, but clumps of endothelium are found scattered, here and there, in the granulation tissue.

Remarks.—This specimen illustrates well the condition of the vessel one month after ligature; the period when an active ingrowth from the external inflammatory tissue is taking place. The walls which have been held in close contact by the ligature have escaped from its control and have unfolded, like an opening flower. In the specimen drawn there is no clot upon the surface, but in the proximal end such a clot rested loosely upon the granulations. The larger vascular spaces, frequently seen in tissue organizing in a thrombus, are here shown to be spaces left between the sprouting granulations, and are clearly distinct from the vessels of the granulation tissue. No communication exists between them as yet. We find in the tissue extending up the sides of the vessels the first indications of a spindle-cell structure, which is destined to be a permanent feature of the cicatrix. The granulations, having grown into the middle segment, are gradually disintegrating and destroying its walls.

The experiments with the double ligature, of which the foregoing are examples, show pretty conclusively that the endothelium is capable of but slight formative power. In the case of active inflammatory changes there was the least perceptible change in this layer of cells, and there was no evidence to show that cells in any number had wandered through the walls of this portion of the vessel. On the other hand, we find its ends open already at the end of a week, and granulation tissue penetrating at these points. At the end of the second week the new tissue has occupied the greater portion of this space, and by the end of the month the lumen is filled with granulation tissue which is now attacking the walls both from without and within. No carmine granules were found within the lumen, although the outer walls had been invaded by masses of granulation-cells. Immigration by wandering cells, as has been already shown, undoubtedly does take place, but has no other significance than to serve as an illustration of the fact that some cells of the invading mass of the granulation tissue may reach the interior of the vessel at an early date. They are few in number at this period, and exercise no essential function in the process of repair which is going on. In specimens carefully prepared with antiseptic precautions, special pains being taken to preserve the vascular supply of nutriment to the endothelium, this delicate cell-layer was found at the end of a week to be in a state of proliferation. The appearance of the cells had changed, some of them appearing as a mere globular form of endothelium, and some as spindle-shaped cells, resembling those drawn in Fig. 3. The presence of other cell-structure than endothelium could not, however, be denied in these cases, for a growth through openings in the elastic lamina, similar to that shown in Fig. 6, was also seen.

By such a series of experiments the fact of the power of the endothelium to proliferate is proved, but it is far from demonstrated that the growth which obliterates vessels or closes wounds in their walls is supplied from this source. The greater portion of investigations bearing upon this question have been conducted at too early a date to show the true relation of endothelium to this process. The later observations in the present series leave no doubt in the mind as to the way in which the process of obliteration has been accomplished. The conditions found to exist in the middle portion of double ligatured vessels differ, in no essential, from those observed in the proximal and distal portions of the vessel, so far as the early stages of the process are concerned; and, if anything is to be inferred from the changes noted there, it is not in favor of the activity of the endothelium as a prominent factor in the process of repair.

FEMORAL. ONE MONTH.

A ligature was placed upon the femoral artery of a dog, and the animal was destroyed thirty-one days after. The wound had healed without any inflammatory disturbance. Before removing the vessel the external iliac was exposed, and the limb injected with Prussian blue. After hardening in alcohol, the specimen was divided longitudinally, longitudinal sections being made through one half, and transverse sections through the other. The ends of the vessel were separated from one another about one quarter of an inch. A portion of the cotton ligature still remained, and was imbedded in the centre of a callus which bound together the two portions of the vessel.

Proximal Portion.—There is apparently a good-sized thrombus distending the vessel in an ampulla-like dilatation; under the microscope, however, this is found to be completely permeated by granulation tissue. The walls of the vessel have separated, and have retracted slightly so that the new tissue is continuous with that of the external callus. Well-formed granulations are found projecting from the thrombus into the free lumen. Just above these the calibre of the vessel is narrowed by a growth from the bruised walls of the vessel. In longitudinal section, the lamina is found to be ruptured on both sides, from which points this growth seems to spring and form in the section a bridge across the vessel just above the thrombus. The tissue of which it is formed grows down the sides of the vessel and becomes continuous with that sprouting from the thrombus (Fig. 19). With high powers the tissue is seen to consist of spindle-shaped cells, and to be directly continuous with the tissue of the media, where some cell-proliferation is also going on. It is covered with a single layer of endothelial cells. At one side there is a small plexus of vasa vasorum, branching out from the media into this bridge of tissue; or, as it proved to be, a diaphragm, which

5

greatly narrows the calibre of the vessel. One or two similar ruptures in the lamina are noticed lower down, whence new tissue is seen growing into the thrombus. The cells found in the granulation tissue are round and spindle-shaped, the latter being quite numerous, and arranged in fasciculi.

Distal Portion.—No thrombus is seen here, but the conditions are otherwise very much the same. The granulations, which could be better studied here, sprout in an irregular manner, leaving spaces communicating with the lumen of the vessel (Fig. 17), which are lined with a delicate endothelium.

The injection mass has run well between the granulations, and penetrates some distance down into the tissue, both at the proximal and distal ends. There is also a rich injection of the vasa vasorum, and the vessels of the external callus; the mass has run particularly well into a plexus of vessels which surrounds the end of the distal portion. These vessels do not appear to communicate with the lumen of the vessel, except at one point, shown in Fig. 19. There is, therefore, a flow of blood into the new tissue from two directions; first, from the lumen between the sprouting granulations; and, secondly from the vasa vasorum and small vessels ramifying in the callus. The latter project some distance into the internal callus, but cannot be followed to a point where they could receive the mass penetrating from the interior. This latter mass indicates a free anastomosis of the spaces between the granulations. (Fig. 17.)

Remarks.—This specimen illustrates the second stage of repair at its height: the opening of the vessel and the ingrowth of granulations. Vascular communication is not yet established between the vessels of the granulation tissue and the lumen.

An examination of the proximal and distal ends of the specimens, ranging from two weeks to one month, shows the development of what may be called the second period in the process of repair; that which follows the separation of the vessel from the ligature, and the expansion of its ends. At the end of two weeks we find, in one specimen, that the walls of the vessel have separated from one another, and that granulation-like masses of tissue are growing into the lumen, partly pushing to one side the thrombus, and partly infiltrating it. Between these granulations are spaces which, lined with endothelium at a latter period, are filled with fresh blood, and, in injected specimens, with injection mass. A careful search fails to establish the fact that these spaces communicate with the capillary vessels which accompany the granulations. By means of these channels, the blood in the lumen of the vessel is able to penetrate some distance into the organizing cicatrix. The development of the cicatrix at this period, although it varies greatly according to the amount of thrombus present, cannot be said to have passed the granulation-stage. No great amount of differentiation of cell-elements has yet taken place; but one begins to observe an accumulation of masses of spindle-shaped cells at certain points.

A careful examination of all specimens within the limits of this period fails to demonstrate any growth which can have been developed from the endothelial layer only. It is true that we find a thickening on the inner walls of the vessel, extending for some distance up above the point of ligature, and that, in earlier stages, a proliferation of the cells of the endothelium has been satisfactorily demonstrated; but there is no evidence to show that this layer plays such an important rôle in the process of cicatrization as has usually been ascribed to it.

In none of the numerous experiments made was an isolated mass of growing endothelium demonstrated. All growths on the interior wall of the vessel were continuous with granulation tissue, or with other layers of the vessel-wall, from which cell-growth was seen to develop. The ruptures of the lamina elastica in the neighborhood were numerous, even when great care had been taken to avoid injury in the application of the ligature. The lamina is, moreover, not a continuous membrane, but is perforated with spaces, leaving a free communication with the tissues of the media. All barriers are, however, at this period broken down, and a superabundance of tissue is provided from without, from which the new tissue, ascribed by authors to a growth from the intima, could easily have sprung. The thrombus, or such portion of it, as has not already been disintegrated and absorbed, is found resting upon the granulations growing up beneath it, and lifted farther into the lumen by them.

FEMORAL. THREE MONTHS.

The femoral of a large Newfoundland dog was tied with a small cotton thread. The wound healed by first intention, and the animal was destroyed at the end of three months. Before removal, the femoral was injected with Berlin blue. The specimen was then placed in alcohol, and subsequently divided into halves by a longitudinal section. The ends of the artery had retracted about an inch, and were untied by a band of connective tissue fibres; all

trace of thrombi had disappeared; the ends of the vessel formed
each a cul-de-sac, the ends of which appeared, to the naked eye, to
be closed by a cicatrix considerably thicker than the vessel-walls,
and showing traces of the injection mass. One half of the proximal end was cut into transverse sections, and longitudinal sections
were made of the other portions.

Proximal End.—(Plate VIII., Fig. 20).—Here the walls of the
vessel are slightly separated from one another, and the intervening
space is filled with a cicatrix, which also occupies a portion of the
lumen, rounding it off, very much like the interior of the small end
of an egg. The surface is lined with a layer of endothelium of normal thickness. Beneath this, growing up on each side, and encircling the fundus also, is a layer of delicate spindle-cells running parallel to one another, and parallel to the arc of the circle formed by the
cicatrix. Under high power, and examined both in cross and longitudinal sections, these are seen to resemble, closely, muscular cells.
The staff-shaped nuclei are characteristically shown. They come
out more clearly and well defined when stained with an aniline dye
mounted in Canada balsam, and examined with a Zeiss oil-immer-
sion (Fig. 22). Prepared in glycerine, they appear as in Fig. 21.
The cells are surrounded by a hyaline, and, at times, a slightly
fibrillated intercellular substance. This band of muscular fibre forms
a crescent-shaped layer, slightly thicker in the middle, with tapering
horns, which present upwards, and line the sides of the vessel for
the distance of about one quarter of an inch from its extremity; one
side being longer than the other. This crescent can be separated
entire with needles; it is found, however, to be firmly attached to
the connective tissue portions of the cicatrix below it, and to the
media at points where the elastic lamina has been ruptured. Bundles of these fibres are deflected, and surround the walls of the larger
vessel, running through the cicatrix. (Fig. 20, f.) Beneath this
muscular layer, and filling out the space remaining between the
slightly separated walls of the vessel, is a mass of connective tissue
fibres such as are usually seen in cicatricial tissue. The fibres
are wavy, swollen, and transparent, running in various directions,
and continuous with fibres of the ligamentous band outside the
vessel.

The cicatrix is pierced in the centre by a large arteriole (v) which
rapidly narrows its calibre, and becomes continuous with the numerous capillary vessels in the new tissue: these anastomose with another rich capillary net-work ramifying in the ligament at its point

of union with the vessel, and ending in two or three small vessels which extend lengthwise through it.

Cross sections of the other half of this portion, made specially with reference to the muscular layer, confirm their spindle-shape, setting at rest any doubts as to there being a layer of flat endothelial cells. Cross sections, including the portion of the cicatrix entirely filling the lumen, again show some of these cells in profile, and if these be examined close to the edges of the bruised media, it can be shown that an outgrowth of the muscular cells of this coat has taken place into the cicatrix. The adjacent cells of the media are also shown in the drawing, with the intervening elastic fibres. (Fig. 23.) Some of these sections include a point where a branch has been given off some distance from the end of the stump. Here the cicatricial tissue (one of the horns of the crescent) is disposed on the opposite side of the vessel. It is thickest there, and as it approaches the branch, tapers away rapidly and disappears at the boundary line. The adventitia as it approaches the end of the vessel is somewhat broken, and terminates abruptly at about the same point as the media. The lamina elastica is well preserved, but broken at one or two points.

Distal End.—About the same conditions exist here as in the proximal end. The layers of the new tissue are not so thick. The muscular layer extends about as far up the wall as in the proximal. The larger vessel, penetrating the cicatrix, is at one corner, and anastomoses with a cluster of capillaries in the fibrous band outside. In all these vessels the injection mass is distinctly seen, and it appears to be supplied from the femoral artery directly and not through the vasa vasorum.

Remarks.—This specimen offers a type of the permanent cicatrix. In shape it is a crescent, the longer horn of which runs up on the side opposite to which a branch is given off, showing the influence of a branch upon the ultimate shape of the cicatrix. (Thoma, Schultz.) The presence of a muscular layer is here clearly established: its presence within the vessel, in what might be regarded as a continuation or thickening of the intima, might give rise to doubt as to the muscular nature of these cells. They are, however, found in the normal intima of arteries occasionally, as well as in veins, and as we have seen, a special formation of muscular tissue occurs at points where extra support is needed, as at the bifurcation of the aorta (Thoma). It seems highly probable, therefore, that we have here a cup-shaped muscle, a sort of levator muscle, whose function is to withstand the pressure of the blood-column.

TEMPORARY LIGATURE.

THREE MONTHS.

A ligature was placed around the carotid of a large dog, tied in a slip-knot with sufficient strength to rupture the inner walls of the vessel, and then removed. On feeling the artery with the thumb and finger, a slight circular indentation could be noticed on the inner walls, although no alteration was perceptible in the outer wall. On dissecting out the vessel a circular indentation was seen, with a slight ridge above and below it. It was placed in alcohol and subsequently divided longitudinally into halves. Both transverse and longitudinal sections were made for microscopic study.

In longitudinal section, two V shaped cuts were found, a short distance from one another, extending through the media, marking the site of the ruptures made by the ligature, which in this case, was adjusted twice.

The cuts are partially filled by cicatricial tissue (Fig. 25), composed of cells similar to those found in the media, and, like them, running in circular bands round the vessel. In cross sections of the vessel they appear as spindle-shaped cells with long nuclei. A few spindle-cells can be seen in the drawing, although the power is not high enough to show them well. The new tissue is intimately connected with the media, and is tolerably distinct from the other layers. A thin covering of endothelium runs over this tissue, and is continuous with the endothelial lining of the vessel, both above and below.

Remarks.—This experiment shows that the new tissue, formed to repair the injury done by a temporary ligature, in all respects resembles that of the media, and is probably a growth of muscular cells. In Zahn's experiments the specimens were examined too early to demonstrate the permanent cicatricial tissue.

In a second experiment of this kind a temporary ligature was placed upon the carotid, and immediately removed, the animal being destroyed at the end of one week. The same kind of V shaped rent was observed, extending through the media to the adventitia. In this case, a thin layer of flat endothelial cells had grown down into the rent on either side, but nowhere formed a thickened layer. The cells of the media are in a state of proliferation for some little distance from the surface of the wound. There are some granulation cells in the adjacent portion of the adventitia. The granulation tissue just beneath the endothelium is evidently formed by a growth from the media, and from cells wandering in from the peri-adventitia, the surface of the wound being already, in great part, covered by an extremely thin layer of endothelium. There could be no better proof of the activity of the media, and of the subordinate part played by the intima than that afforded by these two specimens.

CAROTID. FOUR MONTHS.

This specimen was prepared to show the completed cicatrization, but, on removal it was found that the process was not fully accomplished at the proximal portion. The two ends of the vessel were united by a cord-like mass, and it was evident that the vessel was slowly becoming obliterated up to the first collateral branch. Owing to the length of the main trunk of this vessel, which gives off no branch of size for a considerable space, a much longer time was evidently necessary for a completion of the process of repair, than when the ligature is applied nearer a large branch.

Proximal Portion.—From the point of ligature, for a considerable distance, the vessel appears to retain its form and to be filled with a growth of cicatricial tissue, but, on examination of longitudinal sections, it is found that the walls of the vessel have disappeared and are replaced by a ligamentous band of connective tissue. On nearing the lumen, traces of the media are found, and, a little higher still, the media is seen unaltered, its inner margin being sharply defined by the lamina. Here it encloses a connective tissue, with longitudinal fibres, capillary vessels and numerous round and spindle-shaped cells. The thrombus rests upon this tissue and is still in part preserved, slightly distending the calibre of the vessel. It does not appear to be infiltrated with new tissue, but is encapsuled by it, granulations closing in about it, and separating it from the lumen. There is a thickening of the intima, covered with a new growth of endothelium extending from this point for some distance equally on all sides of the vessel, and gradually tapering off to a normal thickness. It seems as if the clot were of comparatively recent origin, and were assisting in the further obliteration of the vessel.

Distal End.—Here the cicatrization is complete, a point having been reached where a branch is given off (Fig. 26). There is no thrombus. The vessel is converted into a cord-like mass near the point of ligature, and, nearer the point shown in the drawing, is filled, as in the other portion, with connective tissue. That part bordering upon the lumen, however, consists of a tissue abounding in cells quite different in character from those above mentioned. Here we find a tissue, consisting of longitudinal rows of spindle-shaped cells with staff-shaped nuclei which form a cul-de-sac from the apex of which a vessel projects into the new tissue, and breaks up into a capillary net-work. The shape of the cicatrix, as determined by the branch, is characteristic. It does not extend symmetrically up the sides of the vessel, but is thicker and longer on the side opposite to the branch. It is interesting to note that, on the side on which the branch is given off, the cicatrix extends beyond the point of origin of the branch, instead of terminating at that point, as has been represented by observers. (Schultz.) With high powers, the staff-shaped nuclei are well shown (Fig. 28), and it will be observed that at some points shown in the drawing there are breaks in the continuity of the lamina through which similar cells are growing from the media. Cross sections of the vessel at this point show that these cells are not flattened endothelium, but are genuine spindle cells (Fig. 24). The inner surface of the cicatrix is lined with a single layer of endothelium.

Remarks.—This specimen illustrates the manner in which a long trunk is converted into a cord. The process is still going on in the proximal end; the walls of the vessel gradually disappearing as a new growth penetrates deeper into its interior. There is great contraction of the vessel, which is much reduced in size. The peculiar shape of the completed cicatrix, at the distal end, is evidently intended to adapt the size of the tube to the diminished blood-supply which now flows down to be diverted into a collateral branch, or to find its way into a small terminal arteriole. Long branchless trunks, like this, require a much greater amount of time to complete the process of repair than vessels having numerous branches, although they may be equal in size. The growth of muscular cells into the new cicatricial tissue, through the intervals in the elastica, is a reminder of what has been observed by Thoma in the intima of the carotid and aorta of man, and in the obliterating tissue of the umbilical artery.

These specimens show that at least three months are required to complete the process of repair in the large arteries of dogs. The period is, however, a variable one, and appears to depend upon the proximity of a large branch rather than upon the size of the vessel. In the carotid artery, the process of obliteration is not yet complete at one end, but, at the other, the presence of a branch of considerable size necessitates a blood-supply which renders further obliteration in this direction unnecessary. More time would have been required to have closed the calibre at the other end, where no branch was found for a considerable distance from the point of ligature.

The temporary formations, which were so characteristic of the previous specimens, have now disappeared, and we find the two ends of the vessel, in the case of the femoral artery, united by a slender cord. The cord holding the ends of the carotid artery, on the other hand, is still of a thickness corresponding to the size of the vessel. The internal cicatrix has also become more compact, and in the femoral artery, occupies an inconsiderable portion of the lumen of the arterial stump. In the carotid artery, however, that portion which has become permanently cicatrized contains a much longer internal cicatrix. The size of the permanent cicatricial tissue appears, therefore, to vary, occupying a greater portion of the interior of the vessel in larger vessels. The cord which unites the end of the femoral consists of connective tissue: we see no traces of the vessel walls in it. Under the microscope it is seen to be continuous with the adventitial and periadventitial tissue. The vessels of the internal cicatrix ramify in it. It appears to represent the cicatricial remainder of the external callus, which has now entirely disappeared, and has evidently been considerably elongated by a retraction of the two ends of the vessel.

A microscopical study of the internal cicatrix brings out the chief point of interest in these specimens. In the femoral artery, at its distal end, there is a cup-shaped mass of cicatricial tissue. When observed with a moderately high power, it is seen to extend higher on one side of the vessel than the other, and to be asymmetrical. It is pierced at the most dependent portion of the lumen by an arteriole, which rapidly breaks up into smaller vesesls which lose themselves in the cord outside. The centre of this cicatrix is made up of ordinary connective tissue, but this is covered by a layer nearer the interior of the vessel, which at once attracts the attention of the observer, consisting as it does of elongated, spindle-shaped cells running parallel with one another. A thick layer is thus formed beneath the new endothelium, and is prolonged upward on the sides of the vessel in the manner already described. These cells are closely placed together, and arranged with great regularity parallel to the surface of the lumen. When studied with such re-agents as serve best to display their nuclei, these are found to be greatly elongated. In cross sections the contour of the cells is circular. In short, the general appearance resembles closely the muscular cells. Their disposition in the cicatricial tissue, and the sharp contrast which they afford to the connective tissue portion of the cicatrix, also favor this view; as does also their arrangement round the walls of the arteriole which penetrates the cicatrix. A careful comparative study of these cells with those found in the media of the same specimen fails to bring out any essential points of difference; their size, shape, and general bearing towards reagents being the same.

The drawings taken from the four months' specimen of these cells shows not only the arrangement of the nuclei, but also their direct origin from cells in the media. (Fig. 28.) Such slight solutions of continuity as are frequently seen in vessels are probably normal: there are, however, always a number of such openings, some of considerable size near the point of ligature, due to injury received. In the early stages of repair, a growth of cells may be frequently observed penetrating the thrombus through one of these clefts. (Fig. 6.) Later, fully developed muscular cells are found in these clefts. In the specimen of three months, where the walls were ruptured internally by a ligature which was immediately removed, the growth resembles closely newly formed muscular tissue. The cells here are arranged circularly, and not longitudinally, as in the other specimens. The shape of the cicatrix in the permanently ligatured vessel, deserves notice. The asymmetry of the cicatrix in the femoral artery is due to the presence of a branch of considerable size; the long horn of the new tissue running up the side of the vessel opposite to that from which the branch is given off. In the four months' specimen this peculiarity is more strongly marked. There is a large branch near the distal end, the cicatrix forming quite a thick layer opposite that vessel. It is interesting to note that the cicatrix also continues beyond the point of origin of the vessel on the other side. (Fig. 26.) The design of this particular arrangement is apparent. The vessel has contracted at this point and above much within its former calibre; but this has not been sufficient in degree at certain points to adapt the lumen to the greatly diminished blood current, which now has an outlet through the collateral branch and terminal arteriole only. The shape of the cicatricial tissue accurately complements the contraction of the walls by effecting an adaptation to the size of the blood-stream. This point was first noted by Schultz, and its analogy with the narrowing of that portion of the arterial blood-vessels specially concerned in the placental circulation has been traced by Thoma. But Schultz failed to note its continuation on the side of the branch above its point of origin. In exceptional cases, where a trunk exists with no branches except at long intervals, the process of repair may be considerably delayed, owing to the time requisite for a gradual obliteration of a considerable portion of the arterial calibre. In such a case as this, it is probable that portions of the vessel-wall are gradually disintegrated and absorbed during the reparative process, in a manner similar to that which occurs in the middle portions of a double ligatured vessel.

The nature of the cicatricial tissue is complex. The inner surface is lined with an endothelial layer in no way differing from that seen in other portions of the vessel. Beneath this is a layer consisting largely of muscular cells, arranged longitudinally, and giving firm support to the walls of the end of the arterial stump. Outside of this is a layer of connective tissue. We thus have three layers, or coats, to the newly formed wall closely resembling those of the normal arterial wall. The well-marked type of the muscular portion of the cicatrix may, in part, be due to the fact that muscular tissue is unusually abundant in the arteries of dogs.

THE HORSE.

The following series of experiments was performed to show the coarse appearances of the process of repair in arteries. In each case the carotid artery was tied, because it is the most convenient of the large arteries upon which to operate in the horse. No special pains were taken to preserve antisepsis, and some of the specimens were ruined, owing to the degree of suppuration. In those selected, the amount of inflammation was, however, not excessive; and the results obtained were such as served to show prominently the usual changes which the ends of the vessel undergo during the process of repair. A specimen which would show the final cicatrix at its ultimate stage of development was not obtained, but the investigations were carried sufficiently far to illustrate the growth of the callus, its subsequent absorption, and the various alterations which take place in the arterial walls, as well as those observed in the interior of the vessel.

Owing to the great size of the vessels, this series is particularly well adapted to demonstrate to the naked eye the somewhat complicated nature of the process, which is shown to advantage by making longitudinal sections, and dividing the specimen into equal halves, thus preserving all reparative changes in their proper relations to one another. The usual custom of dissecting away all external inflammatory tissue, thus laying bare the ligature, has been carefully avoided.

Two Weeks.

The vessel is imbedded at the point of ligature in a callus about three quarters of an inch in diameter, and about two inches in length. The silk ligature appears to be uninjured. The wound had healed by first intention, and the ligature is completely imbedded in the callus. On either side of the ligature the ends of the vessel, which are still in close contact with it, appear to be distended by the thrombi, particularly the proximal end which has the usual ampulla-like dilatation. There is, in reality, no dilatation, but the vessel, above and below the thrombi, has greatly contracted. The proximal thrombus is shorter, owing to the presence of a branch near by. Why it should be so much lighter in color than the distal thrombus which is dark and stratified, does not appear. They both seem to be intimately connected with the walls of the vessels at their bases only. The walls have not yet opened, the media forming a continuous line above and below the ligature. To the uninitiated observer it would appear that the healing process is complete. There is evidently no appreciable thickening of the intima, certainly not enough to form a thickened layer, although possibly, a cell growth sufficient to hold the base of the thrombus firmly in position, may exist. The adventitia may be traced down the sides of the vessel into the callus (Fig. 1. Frontispiece), but apparently it has been absorbed at the point of ligature.

Remarks.—The first stage of the healing process, namely, the absorption of the outer walls of the vessel by the granulation tissues of the callus, so that they may retract and expand, admitting a growth of granulation tissue, has not yet been completed.

One Month.

A silk ligature was placed on the carotid of a mare, and the animal was destroyed one month later. The wound healed without special irritation, a small sinus still discharging slightly at the time of death. The artery, on being removed, was found enclosed in a very extensive callus, a spindle-shaped mass about five inches in length and one and three quarters inches in thickness. The specimen was accidentally cut open just above the ligature, and some puriform material oozed out. After hardening the specimen and laying it open longitudinally this fluid was found to come from a portion of softened thrombus. In the centre of the callus is a cavity, the size of a pea, partly filled with granular *débris*, indicating the site of the ligature; it communicates by a fistulous opening with the surface. Above and below, and slightly retracted from the liga- ture, the walls of the vessel can be seen completely imbedded in the inflammatory tissue surrounding it, and filling its lumen, which is obliterated, for the space of an inch and a half on either side of the ligature. Beyond this internal growth lies the thrombus, about an inch in length. In the proximal portion there is a mass of softened thrombus at the junction of the clot with the internal growth and this softened mass communicates by a narrow fistulous track with the ligature. There are two such softened spots in the distal portion, one lying in the middle of the internal callus, and both communicat- ing with the ligature, as on the other side. As the artery enters the callus, there is a noticeable thickening of the media easily seen with the naked eye. As the point of ligature is approached, the outline of the wall becomes less distinct, and the media can no longer be traced. This thickening of the media is found to be due to a prolif- eration of the muscular cells of that layer. At its widest part, where the coat is treble its natural thickness, there is a marked increase in the number of cells. In some places several small cells occupy the space of a muscular cell. The muscular cells, which are usually matted together in bundles, are now separated. Both the nuclei and bodies of the cells in cross sections appear enlarged, and, in longi- tudinal sections, they are shorter and more numerous. Nearer the ligature, the elastic lamina disappears, and capillary loops, and a small cell growth may now be seen crossing the media toward the axis of the vessel. Finally, the longitudinal elastic fibres are separated more and more from one another, and the identity of the media is completely lost in the granulation tissue. The intima, above and around the thrombus, shows no special change; in the part occupied by the internal callus no trace of it is to be seen.

Remarks.—We have here an example of unusual inflammatory reaction. A softening of the thrombus has been produced, but this has been compensated for by an increased development of callus, both external and internal. The ends of the vessel have retracted and opened, and an extensive growth of granulation tissue has found its way into the intima, replacing the thrombus as it breaks down. The activity of the cells of the media is specially significant; this is not a simple breaking down of the muscular cells, but an active participation by them in the proliferation. It seems but natural to assume that these cells play some part in the process of cicatrization. This specimen represents the reparative process well advanced in its second stage.

Two Months.

One month after the ligature was applied the wound reopened, and a small sinus continued to discharge until the animal was destroyed. This sinus was found to communicate with the ligature, traces of which could still be seen.

On laying open the vessel by a longitudinal section, we observe that the ends of the vessel have retracted some distance from the ligature, and have opened, allowing granulation tissue to block up the lumen. The external callus is about two and a half inches in length, and one and one half inches wide. The internal callus occupies about three quarters of an inch of the proximal portion, and one half inch of the distal portion. Remains of the thrombus are seen attached to these growths and covering them like a scab. (Fig. 2, Frontispiece.)

In the proximal portion the growth extends much further up the sides of the vessel than it does in the centre, forming a cup-shaped end to the lumen, at the bottom of which the remains of the thrombus are firmly attached. The growth is covered with a thickened layer of endothelium, beneath which are found bundles of spindle-shaped cells with long staff-shaped nuclei closely resembling the muscular cells of the adjacent media. These cells occupy the upper part of the internal growth, and accompany some of the vessels which dip down into it. There are a good many capillary vessels ramifying in a hyaline connective tissue which fills the lower part of the callus. A few vessels come down from the lumen; the communication between the two sets is not yet definitely established. No marked change is seen in the media. The connective tissue layer of the intima can be followed with the other coats of this vessel. The shape and structure of the internal callus of the distal portion are essentially the same as that found in the proximal end. The clot, however, extends more deeply into the new tissue below it. It is here found to be composed of amorphous and more or less fused red-globules and fibrine; the latter being either fibrillated or "canalized." Its meshes are more or less completely permeated with a round cell-growth. The layer of muscular cells is also found in this portion. No communication of the vessels of the granulation tissue with those coming from the lumen is seen.

Remarks.—This specimen affords a typical example of the external and internal callus formation at the height of their development. The expansion of the ends of the vessels shows the ease with which granulations may enter it. The incorrectness of the term "ulceration of the ligature" is exemplified here. There has been no ulceration or cutting action, except at the moment of ligature. The inflammatory process set up, has converted the adventitia included in the ligature into soft granulation tissue, thus enabling the ends of the artery to retract, and leave the ligature completely isolated.

The growth of granulations into the lumen of the vessel underneath the remains of a blood-clot closely resembles that form of reparative process which is usually termed "healing by scabbing."

FOUR MONTHS.

The ligature was applied at about the middle of the common carotid. This experiment was performed in order to procure a specimen of the fully cicatrized artery, but, on removing the vessel, it was discovered that the callus had not yet been entirely absorbed.

The ends of the vessel have separated from one another about an inch, the walls have expanded, admitting a growth of tissue which fills about one half an inch of the vessel on the proximal side of the ligature. This internal growth, on the distal side, is about an inch and a half in length, and some clot is still adherent to it, extending nearly up to the first collateral branch. The ligature had been placed midway between two branches which are about six inches apart.

The internal growth consists of young connective tissue, containing many long spindle-shaped cells with staff-shaped nuclei. These are seen best in those portions which extend up the sides of the vessel for some little distance beyond the central portion of the growth. There is a thickened covering of endothelium over the new tissue. Blood-pigment is found in this tissue in abundance. There are several blood-spaces running down into it from the lumen, so that its free surface presents a very irregular outline. In their minute anatomy, the proximal and distal ends do not materially differ from one another. The vascular connection between the blood-spaces, communicating with the lumen, and the arteries, entering the end of the vessel with the ingrowing tissse, is evidently established. The tissue at the point of its entrance into the vessel, has still the appearance of a granulation tissue. (Fig. 3. Frontispiece.)

Remarks.—This specimen illustrates the very slow progress which a long vessel without branches makes in completing the healing process, the final cicatrix not being formed until the portion intervening between the branches has become obliterated. This is effected by a slow growth of the internal callus along the interior of the vessel, and by the retraction and contraction of the walls and their partial absorption.

This series covers the period beginning with a well-formed callus surrounding the ends of the vessel, still firmly closed by the ligature, and ending with the stage during which the process of absorption of the external callus is far advanced, and the development of the final tissue which is to form the permanent cicatrix, is taking place.

In the first specimen, we see the stage which is usually taken to represent an artery after ligature, and before examining the later specimens, it would appear that the healing process had been completed, and the ends of the vessel-walls firmly moulded into a continuous layer around each end of the vessel; but, a few weeks later, the ends of the vessel have unfolded, and in Fig. 2, of the frontispiece, we see the full development of the second stage of the process of repair. The interior of the vessel has been filled with a mass of inflammatory tissue directly continuous with that formed outside, which has infiltrated and replaced the blood-clot, now largely absorbed. In the first of these specimens, there is no change noticed in the vessel-walls; and so far as we can judge the tissue, found within the laminæ in the second specimen, emanates directly from the granulation tissue. This has melted down the fibres which held the walls together, and has allowed them to retract and to permit an entrance of the active cell-growth to the interior. These facts are obvious, even to the casual observer, and a careful examination with the microscope fails to show that, up to this point, the coats of the vessel have taken any active part in the process. No thickening of the intima can be found to account for the presence of the obliterating tissue. This tissue is continuous with the inner coat, the cells of which may participate freely in the reparative process; but there is no marked outgrowth from the intima. The media is still sharply defined from the surrounding mass of granulation-tissue, but its cells have already begun to proliferate and to prepare for their participation in the formation of the ultimate cicatrix. The tissue of the internal callus projects into the interior in irregular, granulation-like masses which are covered by the remains of the clot, very much as the dry blood of a wound may form the protecting cover of a granulating surface. In the specimen of one month, the inflammation has been sufficiently severe to exaggerate these processes greatly. The external callus forms a spindle-shaped mass of inflammatory tissue nearly half a foot in length; the thrombus is correspondingly elongated, and, at several points, has undergone a puriform softening. The conditions are favorable for a breaking down of the material which plugs the interior of the vessel, but this tendency has been overcome by the excessive callus formation which has penetrated the interior for a considerable distance. The balance between the suppurative and reparative process is thus maintained to an extent sufficient to prevent secondary hemorrhage. In the vessel, tied four months previous to removal, illustrated in drawing 3 of the frontispiece, the temporary formations are in process of absorption; the ends of the vessel have retracted still further from one another, and it can be readily seen from what structures the band of tissue, which unites the two ends of a ligatured artery, is to be developed. The internal growth fills a considerable portion of the interior of the vessel, and it is probable that this growth would have extended still further until the entire trunk had become obliterated throughout the space intervening between two large branches. The final formation of the cicatrix is thus considerably postponed in this case beyond the average time of most arteries, owing to the unusual length of the trunk. A differentiation of the cell-elements of the internal callus is, however, already taking place, and cells, closely resembling muscular cells, can be seen in large numbers. A study of the walls of the vessels involved in the inflammatory changes suggests that an absorption of portions of the various coats occurs during the series of changes. The separation of the two ends of the artery is partly due to this process, and partly to the retraction of the walls.

6

CHAPTER III.

MAN.

LIGATURES IN CONTINUITY.

THE following specimens have been selected to represent the most important stages in the process of repair, and include most of the arterial trunks of largest size. They are taken from the collections of the Army Medical Museum and the Warren Museum of Anatomy, some of them being of considerable age, and some prepared within a period of twenty-five years. Although not presenting the most favorable conditions for microscopic work, the histological elements were sufficiently well defined by careful staining to determine the nature of the various tissues studied. Some of the preparations of hematoxylin gave the best results, although carmine and the aniline dyes were also used, the sections being mounted both in Canada balsam and glycerine.

In all the specimens examined, embracing the period during which the callus is to be found, this structure had been dissected away, showing the vessel either as it would appear directly after application of the ligature; that is, as a hollow cylinder, constricted at one point by a knot, or as two fragments slightly separated from one another, but held together by a narrow cord. Both of these results had been obtained by dissecting away the callus. It needs but a glance at the three specimens of the carotid artery of the horse, shown in the frontispiece, to be convinced of the superiority of the method there employed for displaying the vessel to the best advantage. All results of pathological changes have been preserved, and the specimen is divided into halves by a longitudinal section. In several of the preparations selected for study, the interior of the vessel had never been displayed, and their value was greatly enhanced by treatment in this manner, a portion being obtained at the same time for microscopical examination.

Although the evidence of the existence of a callus had been destroyed by the dissection, the descriptions in some cases gave most

satisfactory testimony to the existence of such a structure. A reference to the description of specimen 1684, Army Medical Museum,
shows that the ends of the vessel at the point of ligature were imbedded in a callus of very considerable size.

COMMON CAROTID ARTERY.

FOUR DAYS.

Specimen No. 1748, Warren Museum.—L. C., Female, æt. 32,
presented herself at the Massachusetts General Hospital in August,
1857, with "a vascular encephaloid tumor of left temple, involving
orbit and cranial cavity." The left common carotid was tied August
8th; the right common carotid was tied August 18th; the patient died
August 2? carotid

The nous throf the right carotid was placed a little below the
point ,ot, no injubn. On laying open the specimen longitudinally,
a large. The me was found below the point of ligature distending
the ve tely separa pulla-like shape near the ligature, and filling about
one h ion. A cell h of the vessel, growing gradually smaller and
being .media; loosed to the walls in its entire length, except at the
small c the proximal pex. It proved to be a hard and extremely dark
colored; to be no c distal thrombus was much smaller, more friable,
and li nk, consists a great deal of it dropping out on opening the
vessel. lumen of the up the internal carotid beyond the end of the
portion lence whatevereserved.

Pro has been dissc At the point of ligature, a low power shows
both th. lia ruptured; the former being torn entirely
throug arks.—The proc about two thirds on the other. The adventitia is · by a suppurat dense and thick bundle of fibres by the
ligature, and does not appear to have been injured. There is no
marked curling up of the broken ends, but they appear frayed
out and are entangled in the clot. The walls are distended by
the thrombus at the point corresponding to its base, and are in
consequence much thinner than normal. The media is especially
thinned.

The intima, which in this vessel has quite a thick connective
tissue layer, terminates on either side in a tapering point at the edge
of the ruptured media. There is no change in this layer near the
ligature; a number of white corpuscles, grouped together at various
points, cling to it, but they seem to belong to the thrombus. A
large number of leucocytes are collected between the ruptured edges

of the media, and are entangled in fibrine-fibres. These together form quite a mass, filling the central third of the thrombus for some distance. The thrombus is irregularly laminated, and has, when seen with a low power, a marbled appearance, the irregular arrangement of the fibrine and white corpuscles, which are always together, breaking the mass of red corpuscles up into lobules of varying size. The clot is exceedingly tough and difficult to cut with the microtome, owing apparently to its density. It terminates in a short pyramidal apex consisting of fibrine and white corpuscles. The walls of the vessel throughout, except at this terminal point, are adherent to it, and, as the thrombus diminishes considerably in ··cumference in its lower half, are much contracted at this point. presear the apex there is a marked thickening of the intima, showʼe moander a high power, a growth of spindle and round cells, whicl the cœct into the clot a short distance, and, when broken away fum of ꞈpresent a serrated edge. Nowhere else is any such chaꞈ prepared w intima found. There is little change seen in the walls ofꞈting the moortion. A small fragment of clot, composed of red corpusꞇtological elꞈ still to the ruptured surface. There is not the sameg to determ'face as upon the proximal side of the ligature, but the ꞈ the preparaꞇs to be thrown into a series of folds radiating from thꞈine and theThe appearances of both sides of the ligature closely ꞈ both in Canꞇ shown in Fig. 2. Following the walls of the vessꞇ   a slight thickening of the intima is observed at tlg the period of the vessel on the sides opposite the angle, a poꞈe had been drtion of this thrombus ends.   ꞇl appear direct

Remarks.—In this specimen all externalꞈow cylinder, corꞈissected away in accordance with the usual methodꞈs slightly separatꞈ It is an interesting fact that no signs of new tissue-ow cord. Both be found within the vessel, except at a point quite remoꞇe ꞈꞈꞈꞈ ꞈꞈ ligature. The growth in the intima near the apex of the thrombus both favors the narrowing of the lumen, and serves to hold the thrombus firmly in place. Is such a growth to be regarded as of traumatic origin, or simply as a formative process intended to adapt the parts to new conditions?

There is evidently a true dilatation of the proximal end of the vessel, for the walls are put upon the stretch, and much thinned; the appearance is not produced by the contraction of the vessel below.

The appearance of the distal portion of the vessel at the point of ligature is the one usually seen. The fact that the ruptured edges of the media are not seen is due to the collapse of the vessel, or rather to the lack of distension of this portion.

COMMON CAROTID ARTERY.

FIFTEEN DAYS.

This vessel was also part of specimen 1748. The ligature had been placed very high on the left common carotid. The proximal thrombus extended nearly to the aortic arch; the distal was very small, being confined to the main trunk only: both thrombi were very friable and unattached, and dropped out on opening the specimen.

Longitudinal sections show that the inner coat, which in the common carotid is of unusual thickness, has not been divided, but is continuous through the ligatured point; beyond a slight bruise at one spot, no injury has been done to it by the application of the ligature. The media and adventitia have, on the other hand, been completely separated, probably by the softening effect of the cell infiltration. A cell-growth is found in the inner layers of the ends of the media; loosening its fibres and dissecting up the intima so that, at the proximal end, this coat is raised up like a vesicle. There appears to be no change in the elements of the intima, which, in this trunk, consists of quite a thick layer of longitudinal fibres. In the lumen of the vessel on either side of the ligature, there is no evidence whatever of any repair. Such external callus as existed has been dissected away in preparing the specimen for the Museum.

Remarks.—The process of repair seems to have been arrested, probably by a suppuration and softening of the external callus. There is insufficient provisional growth to replace the retracting outer coats. The non-adherence of the thrombus is probably due to the lack of rupture of the intima. In the right carotid, it will be remembered that the inner coats were ruptured, and the thrombus firmly attached. Secondary hemorrhage from the distal end could not have been long delayed had the patient lived. It is particularly worthy of note that no tissue is found growing from the intima, although unusual facilities exist for studying that tunic..

The second stage of the process of repair is beginning. The outer walls have released themselves from the ligature, and granulation tissue is about to penetrate the inner coat, which in this case has not been ruptured.

SUBCLAVIAN ARTERY.

FORTY-SIX DAYS.

Specimen 1684, Army Medical Museum.—The specimen is thus described in the catalogue of the Museum :

"A wet preparation of the left subclavian artery, forty-six days after ligation in the third portion, for traumatic aneurism of the axillary, after gunshot, and twenty-eight days after the ligature came away. Capt. T. F. J., "B." 13 Va. Cavalry (Rebel) 31." A carbine ball passed through the brachial plexus of nerves, and cut the axillary artery one and one half inches above its termination. Middleburg, Va., 21st June: admitted hospital, Washington, 23d June; a circumscribed traumatic aneurism at the seat of injury appeared 12 July; subclavian ligated at the external border of the scalenus by surgeon John A. Lidell, U. S. Vols. 14th. Aneurismal sac opened spontaneously 19th July; ligature separated without hemorrhage 1st August; profuse hemorrhage from the sac arrested by injection of solution of persulphate of iron, 6th; hemorrhage recurred 10th, 11th and 18th; died, exhausted with suppuration and hemorrhage, 19th August, 1863."

The point of ligature is represented by a ligamentous band about one quarter of an inch in length, but this is evidently an artificial production, for it is stated by Dr. Lidell, in his account of the case (Surgical History of the War, Part First, p. 544), that the artery on each side of the ligature, "is surrounded by a dense mass of new connective tissue, so thick and dense as to make it a little difficult to get at and remove the specimen without injury." Evidently there was a large external callus which had been dissected away. The vessel was not open, but it was evident that there was a thrombus of considerable size in the proximal portion, which for the space of over an inch, gave off no branch. The specimen was opened in such a way that a longitudinal section was made through both the ends of the vessel without any injury to the connecting band, the parts removed being used for microscopical examination. The lumen of the proximal portion appeared to be occupied by a thrombus about three quarters of an inch in length, filling out the vessel to nearly its full size, and tapering off in an elongated point on the side opposite the origin of the first branch. On the distal side, the process of cicatrization seemed to be complete; a small thread-like fragment of clot still hanging from one portion of the cicatrix.

Proximal End.—On microscopical examination it is seen, that most of the thrombus has become "organized;" that is, has been replaced by new tissue. There is still a laminated clot lying upon this tissue, tapering off on one side, as has been described, where it is firmly held by what appears to be a growth arising from the longitudinal fibres of the intima. The new tissue constituting the internal callus, varies somewhat in appearance: in places it is composed of spindle and stellate cells in a hyaline matrix; at others, of ordinary granulation tissue; at others still, there is a considerable amount of blood pigment, and here and there, are to be found fragments of still unabsorbed clot. The tissue has a cavernous appearance due to the presence of numerous blood-spaces, the borders of which are lined with a delicate endothelium. There are also a number of capillary vessels ramifying in the new tissue; whether these communicate with the blood-spaces or not, cannot be determined. Near the point of ligature, the tissue is in a crumbling state, due probably to suppuration. The coats of the artery have separated here, admitting freely the granulation tissue, which supplies probably the greater portion of the new growth, of which the internal callus consists. It is probable also that some tissue is formed by the intima, a growth from which near the apex of the thrombus has already been mentioned, but the longitudinal fibres can be traced down on either side of the vessel nearly to the point of ligature.

Distal End.—The conditions here are in striking contrast to those described above. The process of repair seems to have been completed. The walls of the vessel, as seen in longitudinal section, are slightly separated, and the space thus left is occupied by a mass of cicatricial tissue which runs symmetrically upon the sides of the vessel for a short distance. It is composed chiefly of spindle and stellate cells, the former running chiefly in bundles parallel to the surface of the cicatrix, the latter being imbedded in a hyaline intercellular substance. The cicatrix is pierced by a central vessel, the route of which can not be definitely ascertained. It probably communicates with capillaries in the granulation tissue, which can still be seen lying beneath the tissue just described, and blocking the mouth of the vessel. The filamentous clot is of about the thickness of a silk ligature, and is one inch in length.

Remarks.—The proximal end of this specimen illustrates the process of repair as seen at about the end of six weeks. The vessel-walls have separated and retracted, and an abundant growth of tissue has penetrated and absorbed the thrombus. What may be called the second stage of the process of repair is here represented at its height of development, a large external and internal callus having been formed. It is probable that, in a large arterial trunk like this, a greater portion of the internal growth of the proximal portion will be permanent, presenting appearances closely resembling those shown in Fig. 27. The distal portion is an example of a nearly completed permanent cicatrix. The early termination of the reparative process, here may be due to the presence of a branch of considerable size, preventing the development of a large thrombus or callus; in other words, of provisional tissues, which would require time for their absorption. The secondary hemorrhage was evidently supplied from collateral sources.

A careful examination shows but a slight growth from the intima. This has been sufficient in the proximal portion to attach the end of the clot firmly to the wall, but it is probable that only an insignificant portion of the cell-structure found in the interior of the vessel at this end has been supplied from this layer.

FEMORAL ARTERY.

FIFTY-FIVE DAYS.

Specimen 3983, Army Medical Museum, is thus described:

"Wet preparation of portion of external iliac, femoral, profunda and anastomotica magna, with the femoral ligated in its continuity for secondary hemorrhage." The injury was a gunshot fracture of the bones of the leg, for which amputation in the upper third of the leg was performed, Dec. 4, 1862. Re-amputation was performed at the junction of the middle and lower thirds of the femur on January 15th; the femoral was tied February 4th, and death occurred March 31st, 1863.

The specimen was so arranged that the relations of the vessel in reference to the point of ligature were not easily made out. The proximal and distal ends were apparently drawn forward into the cicatrix, in such a way as to be nearly parallel to one another. The whole vessel was greatly diminished in calibre, although the distal was larger, at the point of ligature, than the proximal end. An attempt had been made to inject the specimen, which rendered the cutting of microscopical sections difficult. One half of each end was reserved for this purpose, and sections were made longitudinally.

Proximal End.—The vessel is much contracted, and gradually narrows to a somewhat sharp end at the cicatrix, where it is united to the distal portion by a ligamentous band of fibres. The walls of the vessel have slightly separated at the point of ligature, admitting a growth of tissue from without. The amount is slight, and there is a slight thickening of the intima on either side as far as the section goes: there is no thrombus.

Distal Portion.—This end presents characteristics one would have expected to find on the proximal side of the ligature, but it could not be made out from a careful study of the specimen, to be other than the distal portion. The lumen is larger than at the proximal end, and terminates as a cul-de-sac: it is filled by a thrombus about one and one half inch in length. The walls at the end of the vessel are slightly separated, and between them is found some connective tissue and pigment cells, beyond which within the lumen is a cup-shaped mass of hyaline tissue in which are imbedded spindle and stellate cells. There are numerous small vessels in this tissue which are seen to communicate by at least two branches with the vessel. A vessel also runs between the separated walls, to communicate with the external tissue beyond: they are all filled with the material used for the injection. There is calcareous degeneration of the media in both portions of the vessel.

Remarks.—We have here conditions resulting from ligature in an amputation stump, followed by a ligature in continuity. The contraction of the vessel had begun at the time of the first amputation, nearly four months before death. The cicatrix at the distal end is a typical example of the cicatrix from ligature in continuity, but there is also a thickening of the intima which foreshadows the development of the obliterating process in the large vessels of a stump.

EXTERNAL ILIAC ARTERY.

ONE HUNDRED AND THIRTY DAYS.

Specimen 3986, Army Medical Museum.—The Catalogue thus describes it:

"A wet preparation of right and left common, external and internal iliac arteries eighteen weeks after ligature of the right external iliac for traumatic aneurism. The specimen shows the ligated artery diminished to a small cord, and the corresponding internal branch much enlarged. Private J. R. L. ' F ' 10th Georgia (Rebel) 19: ball passed through the right thigh, from front to rear, half an inch below Poupart's ligament, 17th September; admitted hospital with wound closed, but with an aneurismal tumor in groin, Frederick, 27th October; external iliac ligated above the circumflex and epigastric by Assistant-Surgeon R. F. Weir, U. S. Army, 6th November; slight attack of hospital gangrene, 25th November, 1862; an abscess near the cicatrix discharged 2d March; arterial hemorrhage, seven ounces, sac opened, femoral necessarily cut, but without loss of blood; no vessel could be found, and death occurred from previous hemorrhage and shock of operation, March 16th.

An examination of the specimen shows a cordlike mass spring- ing from a small pouch remaining at the angle of division of the external and internal iliac arteries. The femoral artery has been laid open and disconnected from the other end of the cord, so that it was not possible to study the distal end of the vessel. The cica- trix at the proximal end was removed for microscopical examina- tion, longitudinal sections being made through it, and subjected to the double staining of eosine and hematoxylin. The walls of the vessel are slightly separated, admitting between them some of the fibres of the cord: the media becomes here irregular and broken in appearance, and gradually tapers off and is lost in the cord. The latter structure consists of longitudinal fibres intermingled with con- nective tissue cells and pigment granules. Just within the vessel- walls there is a crescentic-shaped mass of hyaline cicatricial tissue, the horns of which taper off symmetrically on either side of the wall a short distance beyond. It consists of a transparent homogeneous intercellular substance in which are found spindle and stellate cells. The spindle-cells are exceedingly long with delicate tapering ends and elongated nuclei: they are found singly and in bands, and are arranged parallel to the walls of the vessel; that is, more or less longitudinally and also circularly. This tissue lies upon, but is dis- tinct from, the tissues of the cord. At one point the lamina elastica is broken, and a number of cells are seen growing from the media into the new cicatricial tissue. A small central vessel is seen break- ing up into one or two capillaries which run in the direction of the cord. The cicatrix is covered with a delicate layer of endothelium. (Frontispiece, Plate II.)

Remarks.—This is a typical example of a cicatrized arterial trunk. It has been converted into a cord of fibrous tissue. The internal cicatrix is a shallow one, contrasting in this respect with that seen in the subclavian artery, and in the common carotid artery, the report of which follows. This is explained probably by the absence of any direct blood pressure upon the spot, such as must have been sustained in those cases. It amounts to a cicatrix in the wall of a large arterial trunk, and not at the end of a cul-de-sac.

The delicate tissue, which forms this cicatrix, is beautifully shown, in spite of the age of the specimen, and there can be little doubt, that many of its elements are of a muscular character. How else can we explain the immunity of such a scar from aneurismal dilatation? The healing process is complete; all provisional structures have disappeared, and the parts are now in the form in which they would have remained permanently. We have in this specimen an indication of the full time required for the completion of the healing process in vessels of the largest size.

COMMON CAROTID ARTERY.

FOUR YEARS.

Specimen No. 1749, Warren Museum. The specimen is an old one, having been in alcohol about fifty years. It came from a patient who was said to have died of phthisis four years after ligature of the common carotid in continuity; for what disease was not stated.

The two ends of the artery are separated widely from one another, a long and slender band of connective tissue connecting them together. The cicatrix in the proximal end is about one inch in length terminating at a point about half an inch from the aorta. It has a pigmented look, and might easily be mistaken for a shrunken thrombus, seen, as it is, through a slit in the side of the vessel. Its surface forms an inverted cone presenting towards the aorta; the apex of the cone is, however, not in the centre of the vessel, but a little on one side; probably the side farthest from the heart. (Fig. 27.)

The slit in the vessel was closed with a stitch and that portion containing the cicatrix was removed entire, for the purpose of making horizontal sections. As will be seen in Fig. 27, the walls are well defined, and, at the end of the vessel, are slightly patulous. Just outside is the tissue of the cord which terminates abruptly at the opening in the vessel. At this point some fragments of the media may be seen, which probably represent portions of the obliterated artery.

Within the vessel is an extremely delicate and spongy tissue, filled with blood-spaces, which communicate with the lumen, and also with capillaries and arterioles, which appear to spring from the large blood-spaces, for they branch in the direction of the cord, and not of the lumen; they are also seen to communicate freely with the blood-channels.

The surface of the cicatrix is covered with a layer of endothelium which may be traced down, into the various spaces.

Just below the surface there is a layer of fibres running parallel
with it, consisting probably of spindle-shaped or muscular cells.
The age of the specimen prevents, however, a satisfactory staining
of this portion. Long muscular cells with glistening tape-like bodies
and elongated nuclei are found in many parts of the cicatricial tissue.
In its deeper portion the character of the tissue differs from that
seen on the surface. We find here stellate anastomosing cells and
round cells lying in clear homogeneous or slightly fibrillated intercellular substance, in which are also found numerous masses of
blood pigment. It is in this part of the cicatrix that the large
cavernous spaces are most numerous. These latter are surrounded
with a thin wall of muscular cells, and lined with endothelium.
Some glistening bands, looking like elastic fibres, run somewhat irregularly in a longitudinal direction. In this neighborhood, near the
termination of the vessel in the cord, the new tissue is much less
vascular, and no large channel of communication with outside capillaries is found.

The longitudinal fibres of the intima can be traced, along the
inner wall of the vessel, into the cicatrix, where they do not lose
themselves, but continue on as far as the media goes, forming a layer
separate and distinct from the cicatricial tissue.

The cicatrix in the distal portion is much shorter and terminates
in a funnel-shaped end. It extends much farther up one side of the
vessel than on the other, in fact, into the external carotid, growing
slightly thicker opposite the first branch given off, probably the
superior thyroid. (Fig. 27.) There is no essential difference in
the histological details of the tissue from that already described in
the proximal portion.

Remarks.—As the result of ligature there has been an almost
complete obliteration of the common carotid artery. This has been
accomplished partly by a filling up of the calibre with cicatricial
tissue, partly by a disintegration and absorption of the walls, and
partly by a retraction of the ends of the vessel from the point of
ligature.

The long cicatrix of the proximal end, with its numerous and
tortuous sinuses, surrounded and supported by muscular cells, is
well calculated to support the great pressure that must have been
brought to bear upon it with every pulsation of the heart. It practically represents a long and tortuous blood-vessel, coiled up in a
small space, and gradually terminating in a capillary system. Had
the ligature been placed on the right common carotid, we should

probably have had a cicatrix more like that found in the external iliac artery, the force of the blood current being diverted into a collateral system of vessels. In the present case we have what may be regarded as a gradual narrowing down of the lumen of the vessel, thus preventing a too abrupt obstruction to the blood-current.

The specimens here described illustrate repair in the first week, at the end of two weeks, at the end of six weeks, at two months, at four months, and at the end of four years. Two of them belong to the first period of the process, two to the second, and two are typical examples of the final cicatrization when the series of changes has terminated by the formation of a tissue which is permanent in character.

A comparison with the series of specimens showing repair in dogs and horses brings out no essential points of difference. The preparation of the common carotid four days after ligature is almost a counterpart of that representing the same period in the dog. Fig. 2 might serve equally well for either. The external granulation tissue represented in the drawing has of course been dissected away in the Museum specimen, but we note in both the ampulla-like dilatation of the proximal end of the vessel and the consequent stretching of the wall, which makes it appear much thinner than that on the distal side of the ligature. In some cases this appeared as a distinct dilatation of the end of the vessel with blood, as if it had been packed into this cul-de-sac with considerable force by the pressure of the blood column. The arterial walls are ruptured internally to about the same extent, that is through the media to its outermost layer. The rupture does not always extend to the adventitia, although, when the knot is firmly tied and then removed, allowing the blood to expand the vessel again, longitudinal sections show a rupture extending to the adventitia, as was evidently the case in Fig. 25. The fibres of the outer coat are collected into a tendon-like band which appears to be sufficiently enduring to resist the pressure of the tightest knot: this is the case in the human specimen, as well as in animals. The inner coat of the common carotid is much thicker in man, and consists, in addition to the endothelium, of a thick layer of longitudinal fibres. Occasionally these escape rupture, as in the case of the carotid of fifteen days. In this specimen both the media and adventitia have retracted slightly, and released themselves from the knot; it is not therefore possible to say whether the media was ruptured by the ligature or not, as the separation of this coat might be due to the solvent action of the granulation cells which have already disposed of the outer coat. It is interesting to observe in connection with the preservation of the intima, and the consequent absence of favorable points of attachment, that the thrombus was not adherent to the wall, and fell out on opening the specimen. In neither of these vessels is there any evidence of repair taking place in the interior of the artery, if we except the growth from the intima at the apex of one thrombus, a point some distance removed from the region of cicatrization. Such a growth seems to serve the purpose merely of attaching the thrombus to the vessel at a point whence fragments might readily be dislodged and carried off into the circulation. There did not appear to be any injury to the intima at this point to account for the cell-growth. Such a cell-proliferation was noted in certain specimens in dogs near the apex of the clot. Although no evidence of cell-proliferation could be detected in the intima near the point of ligature, it is probable that some alteration of the endothelium on the surface had taken place in the specimens of four days, as the base of the thrombus was firmly attached to the vessel-wall. A slight growth, such as is seen in Fig. 3. might easily be overlooked. In the four days' specimen, there is a considerable collection of round cells in the centre of the thrombus, in greater abundance even than is observed in Fig. 2. According to those who believe in the theory of the organization of the thrombus, these would be regarded as the first evidences of the reparative change. So far as can be determined in the present case, they appear to be simply a feature of the process of coagulation. There is no evidence of an immigration of cells such as is observed in Fig. 1, although it is probable that a few cells have found their way into the blood-clot from outside the walls. Such an accumulation of white corpuscles can be observed in any clot about a bruise in the vessel-wall, and has no further significance than to serve as an illustration of the mode of coagulation and a means of attachment of the clot to the inner coat of the vessel.

In the artery tied fifteen days before death, we see already the beginning of the second stage of repair. The granulation tissue which surrounds the point of ligature has infiltrated, and softened the tough mass of elastic fibres, constituting the adventitia, and that coat has already begun to retract, as has also the media. In the present case it is not possible to say whether the external callus has been dissected away, or has undergone a suppurative softening. The loosened walls suggest strongly the probability of a destructive process, and it can readily be seen how favorable the conditions for secondary hemorrhage would be if a firm and well-formed external callus did not cover in the ends of the vessel-walls at the period when they are freed from all attachment to the ligature, and begin to retract and to expand. The traditional twenty-one days brings us very close to this stage of the process. In the light of these investigations, the term " ulceration of the ligature " must be discarded for some such phrase as "a suppuration of the callus;" for the ligature, having once ruptured all but the band of elastic fibres, exercises no further active pressure upon the arterial wall, but encloses the fibres of the adventitia, as a ring does the finger, and merely serves to keep the lumen closed. Further separation of the two portions of the vessel is effected by the action of the granulation cells, which disintegrate and absorb the column of fibres of the adventitia; thus enabling the two fragments to separate slightly, and leave the ligature imbedded in a mass of granulations.

In the next two specimens, the subclavian at the forty-sixth day, and the femoral at the fifty-fifth day, we see the full development of the second stage. The description of Dr. Lidell of the "dense mass of new connective tissues, so thick and dense as to make it a little difficult to get at and remove the specimens without injury," leaves little to be desired as a description of the external callus which surrounded the artery on each side of the ligature. The dissection has so carved the specimen as to represent the two ends united by a ligamentous band, the condition usually found at the full period of cicatrization; but the internal callus has been left intact, and we have an opportunity to examine the character of this structure in the human artery at the end of the second month. The most striking feature is the difference between the growths in the two ends of the vessel. In the proximal portion there is an elongated mass of new formed tissue, filling the lumen for the space of an inch in length, a tissue which is so far differentiated as to make it probable that it is to be retained in nearly its present dimensions as the permanent cicatrix. This view is rendered the more probable by a comparison with the proximal end of the common carotid of four years, where a similar elongated mass of cicatricial tissue is to be found, of which mention will be made presently. On the other hand, the distal portion of the subclavian contains a much smaller cicatrix, but, what is more interesting in this connection, one which has reached already its full development into permanent cicatricial tissue, at a period when the other portions of the specimen are still in the second stage of development. This may be accounted for by the nature of the injury which left the distal portion at the seat of ligature comparatively isolated from the blood current, thus making necessary a smaller amount of cicatricial tissue.

In the femoral, tied in its continuity, after amputation had been performed, we see not only the characteristic cicatricial tissue at the two ends of the vessel, but also a general narrowing of the calibre of the artery throughout those portions examined, both by contraction of the walls and by the development of a growth in the inner wall, a peculiarity characteristic of the vessels of an amputation stump.

The external iliac at four months (Frontispiece, Fig. 4), and the common carotid at four years (Fig. 27), are examples of complete cicatrization. In both of these specimens, the long and branchless trunk has been entirely obliterated. In the iliac artery, a shallow cul-de-sac marks its origin from the common iliac. The depression is filled with the typical cicatricial tissue. The remainder of the external iliac is represented by a bundle of parallel connective tissue fibres containing pigment granules, as indicated in the illustrations. In the common carotid, a large portion of that vessel is also represented by a similar cord of connective tissue, but a portion of the old vessel still remains at each end. In opening these ends, however, we see that their interior has been obliterated by the formation of a mass of cicatricial tissue which fills out the calibre of the vessel. At the aortic end, the segment of vessel, thus obliterated, was over an inch in length, and the new tissue extended nearly to the origin of the vessel from the aortic arch. At the distal end, the cicatrix was short, although it extended a little beyond the point of bifurcation. In both of these large vessels there was therefore, a complete obliteration of the lumen. In the iliac, little more than a depression in the wall of the main trunk remained to indicate its origin.

The large column of blood, which formerly flowed through this great vessel was reduced to a microscopic stream running through the arteriole in the centre of the cicatrix. The residual calibre, if such a term may be used, of the common carotid was, however, considerably greater. The tissue, filling the proximal portion, is of an erectile character, containing numbers of large and tortuous blood spaces; these do not appear to communicate with any terminal branch of size, but form a tortuous coil of blood-spaces, the equivalent of a long narrow, tapering vessel which receives the powerful stream directed against it and distributes it over an elastic surface of considerable extent, thus avoiding the excessive strain which would be brought to bear upon a cicatrix so near the heart. In the proximal end of the subclavian artery, a similar mass of tissue exists, which, although still largely mingled with thrombus, and in an incomplete stage of formation, has already blood spaces, and gives evidence of an eventual development into a cicatrix similar to that existing in the common carotid.

In large trunks, like these, the question naturally arises, have the walls been disintegrated and destroyed, or have they simply retracted; the shrunken ends being obliterated by cicatricial tissue? In the common carotid detached fragments of the wall are seen near the distal and proximal ends, suggesting the occurrence of a process which has terminated in the destruction of considerable portions of the wall (Fig. 27). Comparing this specimen with the carotids of the horse and dog at four months, it seems reasonable to assume that the latter show the beginning of the process, of the end of which, the human specimen is an example, that we have an obliteration of a long stretch of arterial trunk by a growth of tissue in its interior, and that by a process of gradual disintegration and atrophy considerable portions of the wall may disappear. That these may, however, also remain and enclose permanently the obliterating tissue is demonstrated in the main vessel of an amputation stump, soon to be described; but there are no data on which to assume that such a condition exists, even after ligature in continuity, except in the limited extent which is seen in the proximal and distal ends of the common carotid at four years.

The time required for the completion of the process of cicatrization in vessels of the largest size is sufficiently well indicated in the specimens of the subclavian at forty-six days, the femoral at fifty-five days, and the external iliac. In the latter, at the beginning of the fifth month, we find the series of changes thoroughly completed. Already the distal end of the subclavian at the end of six weeks has formed a cicatrix almost as perfect in its histological structure as that seen in the external iliac, but the conditions here, owing to the presence of a traumatic aneurism of the axillary artery, may have favored an unusually rapid development. It is safe to assume that the average period in vessels of largest size varies from three to four months, according to the distance of collateral branches from the point of ligature. In vessels of unusual length, like the left common carotid, this period may be considerably prolonged.

7

AMPUTATION.

The specimens examined were, with one exception, taken from the autopsy table and the dissecting room. Although they do not cover all the periods which it would have been desirable to study, they follow the process through its initial stages until it has assumed a sufficiently definite type to throw light upon the permanent conditions which we find in an amputation stump, the result of an operation performed many years before death. The arteries are taken from the arm, the axilla, the leg, and the thigh, and are of the largest size for those regions. For the purposes of comparison, a specimen of obliterating endarteritis of the anterior tibial artery is included in this series.

LACERATED BRACHIAL ARTERY.

Two Hours.

A man was brought into the Massachusetts General Hospital with compound comminuted fracture of the arm. The brachial artery had been ruptured, and protruded from the surface of the wounded tissues, where it could be seen pulsating, but not bleeding. The vessel was exposed some distance higher up, and a ligature was applied, the portion below the ligature being removed for examination. After hardening the specimen, longitudinal sections were made. A considerable portion of the specimen was found to consist of the sheath of the vessel filled with coagulated blood, the artery having retracted some distance.

The sections show that the vessel has been torn obliquely, and that the walls of one side are curled up, leaving a small opening on the opposite side near its extremity (Fig. 4). Both in the sheath below the vessel and opposite the opening there is coagulated blood, and also in the lumen of the vessel. This clot is not stratified. In and near the opening inside the vessel are a few small clumps of coagulated fibrine, and scattered about in the clot are a few white corpuscles. The latter are more numerous in the sheath clot (the "couvercle" of Petit), near the end of the vessel. The laceration has evidently torn the vessel high up in its sheath, so that it retracted deeply into the sheath, which was soon plugged with a firm coagulum. The injury does not resemble that produced by torsion. Hemorrhage has been stopped, not by a twisting of the fibres of the adventitia about the end of the vessel, but by a rapid formation of clot in the sheath, which has subsequently extended into the vessel.

FEMORAL ARTERY.

TWENTY HOURS.

The patient was brought into the Massachusetts General Hospital, having sustained a severe injury, which required amputation of the thigh at the junction of the middle and upper third. The patient died during the night from the shock.

The vessel is an unusually large one. Immediately above the ligature the walls are thrown into folds so that in some sections they are in apposition, and in others irregularly distended. It is not filled out with clot close down to the ligature, but a thrombus about one half an inch in length fully distends the vessel immediately above, a slight prolongation, extending up one side of the vessel about one third of an inch further. It is a mixed thrombus, there being large numbers of white corpuscles, and an abundant deposit of layers of coagulated fibrine. The inner coats are cut at the point of ligature, but they do not curve upwards, and a smooth surface is presented to the base of the thrombus.

Remarks.—The specimen shows a thrombus already forming in a vessel, the interior of which presents no irregularities for fibrine or white corpuscles to adhere to readily, and in an individual whose circulation was enfeebled by severe shock. My notes do not state the fact, but it is probable that the usual antiseptic precautions were observed. Silk thread was the material used for ligature. No valve-like curling of the inner walls exists.

TIBIAL ARTERY.

ONE WEEK.

The vessel was taken from beneath the granulating surface of an amputation stump about the middle of the leg. The media is strongly inverted, apparently in its whole thickness, and, more or less completely, plugs the vessel. The adventitia is already considerably infiltrated with round cells, a mass of which surrounds the end of the vessel and the silk ligature. A bright red unstratified thrombus extends about one and one half inch up the vessel. It is difficult to say whether there has been an actual thickening of the intima. At one or two points this appears to be the case, but these points are not in close proximity to the end of the vessel. There are large numbers of round cells between the surface of the thrombus and this layer, and they adhere to the latter when the two are separated. On the whole, it seems probable that no marked change in the intima has as yet occurred. At the stump of the vessel all wall-outlines are obliterated by the contortions of the media, between the folds of which numerous wandering cells appear to be making their way from without; there is also some infiltration of the media by these cells.

A second tibial of the same date was removed from an amputation stump at about the middle of the leg. Longitudinal sections were made of the half of the specimen nearest the ligature, and cross sections were made of the other half. The media is not folded in as in the previous case, but the two walls lie nearly in apposition, the line of separation being marked by a row of small round cells making their way in from without, but not yet having accumulated in the base of the thrombus. There is some infiltration also of the broken walls of the artery by these cells. The thrombus resembles that found in the other specimen. The noticeable feature of this specimen is a slight thickening of the intima which consists of cells with round and oval nuclei. From one or two points, this growth projects in a columnar shaped mass into the clot for a short distance: this point is some distance from the ligature. The thickening, however, extends over the whole surface of the vessel examined, except near the ligature, where it disappears. In one place there is a rupture of the elastic lamina, and a growth of cells appears to come from the media into the intima.

Remark.—The appearances noted in the vessel of an amputation stump at this period suggest the beginning of an obliteration of the lumen of the remaining portion of the artery, rather than of a healing process limited to the end of the vessel.

AXILLARY ARTERY.

ONE WEEK.

The specimen was obtained from a hospital patient who died after amputation for gangrene following a compound comminuted fracture. The axilla was found in a putrid state at the autopsy. One large vessel was found exposed, but plugged with a clot protruding from its end. Another larger vessel, either the lower portion of the axillary or the beginning of the brachial, was removed for examination. The end of the vessel was enclosed in the granulation tissue, the ligature having come away.

Longitudinal sections show a small clot of pale color, adherent to the end of the lumen. Under the microscope this is seen resting upon a growth of new tissue consisting chiefly of spindle-shaped cells, and also of round cells. It has the appearance of granulations, which are infiltrating the thrombus, growing chiefly in columnar masses. The walls of the vessel have separated slightly, and there is an extension of the growth of granulation cells from the outside into the interior. There is an excellent opportunity to study the intima in this specimen, as the clot is small, and the internal tunic possesses quite a deep thick layer of longitudinal fibres, as is usually the case in large arterial trunks. Curiously enough, the intima is the thickest at the point most remote from the ligature. Here are seen the longitudinal connective tissue fibres, covered with a single layer of endothelial cells, and interspersed with a few cells, having round and oval nuclei. Farther down, this coat gradually diminishes in thickness until but one or two fibres intervene between the endothelium and the lamina: from this point it becomes slightly thicker, and there are a few more nuclei to be seen. In the immediate neighborhood of the ligature it widens out considerably, and becomes lost in the new growth. The endothelium appears to undergo no change, but to be continued out on the surface of the columnar granulations of this new tissue.

Remarks.—On account of the acute septic inflammation, the ligature has separated, and the walls of the vessel have opened sufficiently to admit a growth of granulation tissue which has permeated the base of the clot. This new tissue or " plastic clot " appears to be directly continuous with the intima. That coat, however, probably participates in the growth only to a limited extent, the main portion being derived from the granulation tissue, which has grown in from without. The changes noted in the intima do not indicate, as yet, any material activity of that tunic. The narrowing of the coat above described is evidently an anatomical condition. The slight thickening noted lower down may indicate the beginning of a pathological thickening. The small size of the thrombus is due, probably, to the close proximity of a branch, (the one seen' protruding from the granulations), its friable condition being caused by the septic inflammation. The end of the vessel is, however, for the time being, held together by the granulation tissue, which has found its way in at an earlier period than usual.

TIBIAL ARTERY.
THREE WEEKS.

J. L. was brought into the Massachusetts General Hospital with a compound comminuted fracture of both legs. Double amputation was performed, the patient dying three weeks later of pyemia. Amputation had been performed on one side, about the middle of the leg, and on the other, at the end of the upper third. The anterior tibial, tied at about its middle point, presented the following appearance.

The end of the vessel is surrounded with granulation tissue, in which the silk ligature can still be seen imbedded. Longitudinal sections show that the media has been curled in and has blocked up the end of the vessel, so that only a small number of granulation cells have found access to the interior. The cavity of the vessel is filled with unstratified clot, into which a growth of granulations is taking place, partly filling the vessel for about one quarter of an inch from its end. It seems to spring from the point of ligature, but is adherent at one or two points to the intima, in which coat a striking change is apparent. This change consists in a considerable thickening of this tunic, which greatly narrows the already contracted vessel. The tissue of which it is composed consists of a more or less transparent intercellular substance, in which are imbedded long spindle-shaped cells with staff-shaped nuclei (Fig. 5). Near the surface the nuclei are shorter and more irregular in shape and appearance. Here and there clusters of leucocytes are found attached to the surface. The spindle-shape is confirmed by cross sections, and sections, taken so as to see the cells in face, as well as in profile. No new vessels are noticed in this layer, nor is any distinct communication between it and the middle coat made out. The growth extends beyond that part of the vessel removed for examination, and down to the point of ligature, being reflected upon the curled-in edges of the media. It is in communication with the granulations in the thrombus, the cells of which are similar.

Remarks.—The conditions observed here are those which probably prevail in vessels of amputation stumps, a preliminary stage of the obliterating endarteritis, by which the vessel is reduced to a cord containing a greatly diminished lumen, or one or two small lumina, in order to adapt it to the greatly diminished blood supply to the part. The growth seen in the intima probably comes from its deeper layers, from some portions of the media, and from the tissue accumulating at the point of ligature. It appears, however, to be chiefly derived from the deep layer of the intima. It is probable that this thickening could have been traced to the origin of the tibial, and possibly into the popliteal artery. The shape of the nuclei suggests a strong resemblance to the muscular cells, although at this stage of development, it would be impossible to express an opinion as to their true nature.

TIBIAL ARTERY.

THREE WEEKS.

The anterior tibial artery, at about the junction of the middle and upper third was found protruding from the granulations, in the case mentioned above, the walls being separated, and a large clot plugging the expanded mouth of the vessel.

On dividing the vessel longitudinally the clot is found to be quite short, terminating at the level of a branch of considerable size; the surface which it presents to the lumen is slightly concave (Fig. 14). It is firmly attached to the walls of the vessel, which at this point are undergoing infiltration with wandering cells, many of which have penetrated the thrombus through the media. It is also securely attached to a growth of the intima, which is to be found on one side of the vessel only, that opposite to the branch. Starting from a point somewhat below the level of the upper end of the thrombus, it can be traced to the end of the specimen. At a point a little above that represented in the drawing it has widened, to such an extent, as nearly to close the lumen of the vessel. There is also a slight thickening of the intima on the other side, running down into the branch. The tissue, of which this consists, is of a hyaline, mucous character, and contains spindle, oval, and round cells: a number of young capillary vessels may be seen ramifying in it (Fig. 15). At certain points there are clusters of round cells, and at corresponding points in the adventitia there may also be seen large clusters of the same cells, making their way into and through the media. The upper layer of the intima, consisting of double laminæ, with intervening cells, may be seen lying below this thickening of the intima. The cells, lying between the laminæ, are arranged longitudinally, and have staff-shaped nuclei. This anatomical arrangement is peculiar to the tibial artery. At one or two points on the surface of the new tissue is an appearance of a formation of a new elastic lamina. The absence at certain points of one of the deep laminæ, and the evidence seen at others, of cell-growth taking place between them, as studied in cross sections, suggest strongly the displacement of one of these membranes, as an explanation of this phenomenon.

Remarks.—We have here a condition preliminary to secondary hemorrhage, and an illustration of nature's effort to prevent it. The softening effects of the suppurating granulations have cast off the ligature, and have not provided a substitute for it: the walls have consequently expanded, and the thrombus is protruded. The same intense inflammatory action has, however, infiltrated the walls to such an extent as to hold the thrombus in contact with the vessel so long as it shall hold itself together. In the meantime, the artery is rapidly becoming diminished in calibre in its whole length. The cells seen between the laminæ of the tibial artery are usually regarded as connective tissue cells, but they have much more the appearance of a bundle of longitudinal muscular fibres. The question of a new formation of an elastic lamina has been frequently raised. With this exception, no appearances suggesting such a growth have been observed, and in this case the formation of new elastic membrane is quite doubtful. The appearances of the clot and its shape and relation to the walls are similar to those described by Kocher as seen after acupressure.

POSTERIOR TIBIAL ARTERY.
THREE WEEKS.

One of the posterior tibial arteries in the above case was preserved for examination. The coats of the vessel were folded in at the point of ligature, and the convolutions were glued together by a growth of cells. There were a number of immigrant cells at this end of the vessel. There was a slight thickening of the intima on one side.

FEMORAL ARTERY.
FIFTEEN YEARS.

The specimen was taken from the thigh of a dissecting room subject. Little was known of the man's history: he had been in the almshouse for fifteen years, having lost the leg from an injury received at sea. Amputation had been performed through the junction of the middle and lower thirds. On dissection, a round and firm cord-like mass was found running from the origin of the profunda to the cicatricial tissue of the stump, gradually diminishing in size towards its terminus. The femoral, above this point, was much contracted. Both longitudinal and cross sections were made at various points. These showed that, although apparently a cord, the walls of the vessel still remained, more or less perfectly preserved, and that the lumen had been greatly diminished in calibre by an "obliterating endarteritis."

In longitudinal sections taken near the origin of a larger branch, presumably the profunda, the vessel is found filled with a new growth, which extends some distance above the origin of the latter vessel on the opposite side. Below this point it apparently fills the greater portion of the vessel, but in the centre a tortuous lumen can be traced. Owing to cadaveric changes, the tissues do not stain easily, but a careful study of the different forms of tissue found shows them to be of such a character as to constitute a tolerably complete new wall to the new lumen (Fig. 33). No endothelium can be made out, but next to the lumen is a dense glistening mass of fibrous-like tissue, in which a few staff-shaped nuclei have taken the staining. These cells are seen in greater abundance farther out, and run both circularly and longitudinally. The longitudinal muscular cells are very abundant in some sections. There is also a loose areolar tissue between them and the old muscular coat. This coat has undergone extensive calcification. In cross sections, taken a little lower down, the vessel appears to have a larger lumen. There seems to be, just inside of the old muscular coat, a musculo-elastic layer; still further in, some of the longitudinal muscular cells are now seen in cross section, and finally, next to the lumen, fibrous tissue exists in which a few cells can be found. The lumen is, in most sections, divided into two channels by a band of fibres running across the vessel. In longitudinal sections, taken at the terminus of the cord, traces only of a lumen and vessel-walls are found. The lumen has grown so small as finally to be little larger than an arteriole, and a mass of anastomosing capillaries and arterioles, surrounding this, indicates the system of vessels, in which the artery finally loses itself.

Remarks.—We have here a condition closely resembling what is usually described as "endarteritis obliterans." It would, however, be more appropriately termed "compensatory endarteritis," although not corresponding closely with the changes observed by Thoma in that process. We see here an adaptation to the very greatly diminished blood supply of the limb. We find no distinct cicatrix at the point of ligature, but, instead, a gradual diminution of its calibre in the entire length of the vessel, and a partial obliteration below the profunda. The section shows apparently a vessel of

large size terminating in the cicatrix of the stump, but microscopical
examination shows a diminution of the calibre of the main trunk to
the size of many of its branches. The condition of the new circulation in the stump would be best demonstrated by a corrosion preparation which would represent the new cavities of the vessels without
their covering. This would show that the main trunk soon breaks
up into a number of branches, of nearly equal size, which distribute
the blood to the different portions of the stump.

ENDARTERITIS OBLITERANS TIBLÆ.

W. E. H., 50 years old, entered the Hospital, March, 1884, with
gangrene of the great toe and a portion of the same foot. He had
injured it two months before in very cold weather. The leg was
amputated a few days later at the point of election. The patient
recovered from the operation, and the wound healed well during the
first two weeks, but he died on the seventeenth day, after three days
of severe illness. The following is a summary of the autopsy:
"Calcification of the tibial arteries; chronic endarteritis with obliteration of the cœliac axis; obliteration of splenic artery; traumatic
thrombosis of femoral vein; embolism of pulmonary artery; thrombosis from stagnation in splenic vein; continued thrombosis of portal
vein; marantic thrombosis of spermatic and renal veins; anemic
necrosis of spleen; anemic necrosis of liver." A portion of the
anterior tibial artery was preserved for microscopical examination.
The lime salts having been dissolved out, longitudinal and transverse sections were made.

In longitudinal sections the interior appears to be filled with a
hyaline, homogeneous, slightly fibrillated, intercellular substance
containing round and spindle-shaped cells, the latter having staff-
shaped nuclei and closely resembling muscular cells. Vessels of
considerable size run lengthwise, in a more or less tortuous course:
in some sections there are numerous coils of vessels, giving the tissue
a cavernous appearance. In transverse sections the new tissue is
found to be pierced by two vessels of considerable size, the walls
of which contain, not only a delicate endothelium and several rows
of muscular cells, but also, between these layers, a well defined line
which might pass for an elastic lamina. These vessels are supported
by a connective tissue which fills out the lumen, except at its periphery, where there appears to be a layer of hyaline tissue containing
but few cells, and nearly encircling the interior of the vessel. There
are numerous pigment granules scattered here and there in the new
tissue. (Fig. 35.)

In the media of the tibial artery are numerous calcareous patches, which occupy the greater portion of this layer. This coat is thrown into both longitudinal and circular folds, at times bulging deeply into the obliterating tissue. The lamina is frequently wanting, and a proliferating mass appears to grow into this tissue. A portion of the muscular wall is intact, but in many places, granulation cells are seen in great numbers, and numerous capillary vessels, and even arterioles. There is no appreciable change in the adventitia.

Remarks.—The appearances described strongly suggest an active participation of the middle coat in the obliterating process. The calcification, possibly the original disease, is situated in this coat. Numerous cells strongly resembling muscular cells are seen in the new tissue, and the new large vessels are abundantly supplied with muscular fibre. In fact, a new tibial channel has been constructed. There is a striking resemblance in this process to that seen in the large arteries of stumps, and the similarity of the two conditions suggests the query: may not the obliteration in the present case be an adaptation to, rather than the cause of, the diminished blood supply to the extremity?

In the first specimen of this series we have an example of nature's method of arresting hemorrhage. The injury was one of those severe lacerations for which amputation was necessary. The vessel was seen protruding from the tissue as a solid cord-like mass, an inch or more in length, moving visibly with each pulsation. The vessel had retracted within its sheath for a considerable distance, much greater than is shown in the drawing (Fig. 4), and had been plugged by an internal, as well as by an external thrombus. Other sections showed a slight fissure on the side opposite to the opening seen in the drawing, but the one represented in the figure is the only one of any size, as indicated also by the extravasation of blood between the vessel and its sheath in the immediate neighborhood.

An interesting feature is the absence of the conditions which one would expect to find in torsion, due possibly to the fact that the walls have been torn unequally, and that the opening appears upon one side, rather than at the end of the vessel. There is no curling in of the inner coats, and the adventitia is not twisted about the end of the vessel. Hemorrhage appears to have been arrested by a coagulation of blood within the sheath, followed by a formation of a thrombus inside of the vessel.

In this case there is no stratification of the thrombus, a peculiarity which appears to exist only when the formation of the thrombus has taken place slowly. Such is the case in the next specimen, where, at the end of twenty hours, a comparatively small thrombus has been developed. This may have been due to the presence of a branch, but it is probable that antiseptic precautions had an influence upon its size. This condition of the clot was also seen chiefly in large vessels, in which it was universal to a greater or less degree; it was also noticed in some of the experiments on the arteries of dogs. It is interesting to note here, that, in the axillary artery, the thrombus was exceedingly small, in spite of an intensely septic wound, show-ing that such conditions do not necessarily produce an extensive thrombosis in the human subject.

In the axillary artery the end of the vessel had already begun to open at the end of a week sufficiently to admit a column of granula-tion cells, which were beginning to invade the lumen. In the same wound, a vessel had opened to such an extent that the thrombus was exposed on the surface, the destructive process having prevented the granulations from filling the gap.

The most important observation made on the series of vessels included within the period of the first three weeks, was the growth of tissue on the inner wall of the vessels. The first indications of such a growth were observed at the end of the first week, and at the third week it had already become a layer of sufficient thickness to diminish perceptibly the lumen.

In Fig. 5 we see the character of this growth, which consists of spindle-shaped cells with elongated nuclei. In the tibial arteries, in which this growth was studied, there exist two laminæ, between which similar cells are seen, and the conclusions drawn from the examinations of both longitudinal and cross sections, point to a growth, not only from the inner, but also from the middle coat. This new tissue was found, not only on the inner surface of the vessel, but also projecting in granulation-like masses into the throm-bus, and it was evident that the process, known as obliterating end-arteritis, was to be seen here in its earliest stages.

In most of the tibial vessels examined, the middle coat had been turned upwards into the lumen at the moment of ligature, and had effectually blocked up the opening. Such a valve-like closure was not seen in any of the other vessels examined, and it seemed proba-ble that this incurvation of the inner walls was due to the peculiar character of the walls of this artery. In most of the vessels ex-amined, whether tied in continuity or in the amputation stump, in man, or in animals, there was no valve-like folding-in of the walls, except in the tibial artery; in all other cases such a condition as is represented in Plate I. prevailed.

The condition of the tibial artery at three weeks, with a protruding thrombus (Plate V.) is of unusual interest as illustrating, not only the mode of closure of one of these vessels, but also nature's mode of dealing with an emergency. Owing to the degree of suppuration, the retracting walls have not been sealed by the usual growth of granulation tissue, and have expanded, allowing the thrombus to protrude slightly. The walls on either side, have, however, become infiltrated with wandering cells, which have penetrated the thrombus in large numbers, thus attaching the thrombus firmly to the vessel-wall, and enabling it to resist the expulsive efforts of the blood-column. The blood-stream has, however, been considerably diminished in size and force, by the narrowing of the lumen, which has resulted from an active growth from the vessel-walls, and this growth, has, curiously enough, been most vigorous a short distance above the point of bifurcation of a branch. There the lumen had already become so narrow that it is highly probable that, had the thrombus become softened by a puriform infiltration, hemorrhage might have been prevented by closure at this point. If, on the other hand, healing had taken place, the artery would have probably narrowed at the same point into a fine arteriole, which would have communicated with the capillary system of the stump. The extra development at the point mentioned, the upper portion of Fig. 14, seems to be in part due to a wandering of cells through the walls. This may, however, have been an accidental feature of the severe inflammation which simply stimulated a more vigorous growth from the vessel-walls at a point most favorable for it. This growth has as yet taken on no typical development, but is a myxomatous tissue in a state of active formation. The relations of the growth to the direction of the blood-stream, and the influence exerted upon it by the presence of a branch, are also well shown.

In laying open the thigh, through the lower third of which amputation had been performed many years before, the femoral artery could be traced from its origin to the cicatrix at the end of the stump. At Poupart's ligament its calibre had already greatly diminished, and, on opening the vessel from above downwards, the old lumen was seen to be obliterated a short distance below the profunda. Below this point the condition closely resembled that seen in an obliterating endarteritis. Externally the old vessel appeared to exist as before, though smaller in size. Internally, there was, however, a newly formed mass of tissue, through which a small stream flowed in two or more channels. (Fig. 33.)

The specimen taken from a case of true endarteritis obliterans shows more satisfactorily the histological details. (Fig 53.) As the amputation specimen was taken from a dissecting-room subject, the cadaveric change has injured the tissues sufficiently to interfere with a staining suitable for study with high powers. It is worthy of note that we have here a trunk obliterated through a large portion of its extent, without an absorption of its walls, such as was assumed to have taken place in the ligature of the common carotid in its continuity. The condition of the blood supply to the neighboring parts was not, however, the same in the two cases. In the ligature in continuity the main trunk had been relieved from further duty by an abundant collateral circulation, but, in the amputation stump, the artery beyond the profunda had been reduced in size, but not obliterated, as the blood supply to the part had been permanently diminished, and the need of a powerful collateral supply did not exist.

If we may judge from the specimens examined, it is evident that the course of the process of repair of arteries in amputation stumps is quite different from that occurring after ligature in continuity. In the latter case we have a distinct cicatrix developed at the end of a vascular cul-de-sac; in the former, however, no cicatrix, as such, exists. We see the main trunk, soon after entering the limb, breaking up into a spray of smaller vessels, which distribute themselves more or less equally to the different portions of the stump.

CHAPTER IV.

THE CLOSURE OF THE FŒTAL VESSELS.

THE DUCTUS ARTERIOSUS.

THE obliteration of two important blood-vessels at the time of birth affords examples of nature's method of forming a cicatrix when unaided by surgical art. The analogy between the closure of the ductus arteriosus and the obliteration of a vessel, after ligature in continuity, is one that suggests itself even on superficial examination. A large blood channel is suddenly obstructed and becomes permanently closed at the point of obstruction, and a collateral circulation is immediately established. Here we have an opportunity to examine the nature of the cicatrix in the aortic wall, a cicatrix the conditions of which differ but slightly from those studied, for instance, in the common iliac, at the point of origin of the external iliac artery obliterated by ligature. In the hypogastric and umbilical arteries a large portion of an arterial trunk is removed, cicatrization taking place in the remaining portion; conditions closely resembling those existing in the stump of an amputated limb. It remains to be seen then what analogies exist between the healing of these fœtal structures and the two modes of repair after ligature which have just been studied.

A review of the literature bearing upon the anatomical structure of the ductus arteriosus shows considerable diversity of opinion upon the nature of its histological elements; and existing descriptions of the series of changes which take place during the process of obliteration, are not satisfactory. There is little also to be found, bearing upon the changes in the hypogastric artery, suitable for our present purpose.

The following selections have been made, as they present some points of interest in this connection. For the literature on this subject the reader is referred, however, to the bibliography.

Thoma gives the following description of the anatomy and changes in the ductus arteriosus. In the intima he finds a tissue composed of elastic fibres interspersed with muscular cells, resting upon the elastic lamina, and covered by endothelium, a peculiar ar_

rangement which also obtains in the pulmonary artery near the opening, as well as in the aorta at the same point.　This arrangement is
similar to that already described at the points of bifurcation of the
large vessels from the aorta.　The elastic lamina is continuous with
that seen in the pulmonary artery, and runs nearly to the aortic end
where it is lost.　The media differs materially from that of the adjacent large vessels.　The strong elastic membranes are replaced by
thin narrow membranes and elastic fibres.　There are large numbers
of muscular cells, and the arrangement is longitudinal rather than
circular.　A contraction of these fibres near the centre of the duct
causes that portion of the lumen to be narrower than elsewhere.
At the point where the elastic lamina is wanting, the elements of the
two coats of the ductus and the aorta become mingled with one
another.　The muscular elastic layer of the intima of the aorta is
most pronounced just below the opening of the ductus, and is due to
the fact that the fibres of this tube spread out chiefly downwards
through the media of the aorta, some losing themselves in that layer,
others penetrating the intima aortæ.　The elastic lamina of the aorta,
just at this point, is not a continuous membrane, and the elements
of the two coats intercross somewhat.　This thickening of the intima
disappears a few millimetres below the opening of the ductus.　A
few days after birth the endothelium of the intima of the ductus is
thrown into thick folds, by the contraction in part of its own muscular cells, and in part of those of the media.　They appear also to
be somewhat swollen, and help, with the muscular contraction of
the media, to close the lumen of the vessel.　The intima of the
aorta also undergoes a change which extends from the opening of
the ductus throughout the whole length of the aorta.　It consists
in the development of a hyaline connective tissue just beneath the
endothelium.　Its cells are quite different from the muscular cells
already described, and there are no elastic fibres in it.

As seen at the fifth year the muscular fibres of the ductus have
atrophied and disappeared, and the opening has been closed by a
growth of connective tissue.　In specimens examined, ranging from
twelve to twenty-four years after birth, remains of the fibres of the
ductus are still found in the aorta, some looking like cicatricial tissue, others having undergone the characteristic hyaline degeneration.　It is worthy of mention that Thoma states that the muscular
tissue and the connective tissue of the intima of the aorta closely
resemble one another, and sometimes cannot be distinguished from
one another.

In the umbilical (hypogastric) artery, according to the same author, the elastic lamina is continued only through the pelvic portion of that vessel: from that point to the navel the lining membrane consists solely of endothelium.   At the time of birth there is a contraction of the media which continues back into portions of the internal iliac artery.   There is also a growth of the same hyaline connective tissue which has extended down to this point from the aorta.   A few years later these changes have reduced the opening in this vessel to a short and narrow tube.   Later, all that is seen is newly formed sub-endothelial elastic membrane, lining the remaining cavity of the vessel, surrounded by circular muscular fibres, all of which structure has been developed in the new tissue which closed the lumen of the vessel.*   According to Baumgarten the umbilical (hypogastric) artery in the adult is represented by a thick, hard, round cord, one and one half to three inches in length, to which is attached a more delicate bundle of fibres connecting it with the umbilicus.   This cord contains a canal, which communicates with the superior vesical artery, and is regarded as the remains of an imperfectly obliterated vessel. The old wall of the vessel can be made out throughout all this part, except at the upper end, where the media and the adventitia are somewhat obscure, but the elastic membrane can still be seen.   The extreme end is closed by fibrous tissue.   With the exception of a small fragment, just within the abdomen, at the umbilical end, the vessel is not obliterated, according to the old view.   The process of partial obliteration is complete at the end of six or eight weeks; at the end of four or five months the arteries are attached to the navel by a short band of fibres, which contain no trace of the vessel.   The band of fibres, which attaches the superior vesical artery to the umbilicus in adult life, he concludes, is a band of cicatricial fibres, which have stretched with the growth of the individual, the umbilical artery being actually about the same length that it was at birth. This band resembles the cord which unites the two ends of a vessel ligatured in continuity which have retracted and stretched apart. The ductus arteriosus is an example of what he calls the " typical process of incomplete obliteration."

In order that the following series of reports may be understood, a word of explanation is necessary.   The ductus was re-

---

* "Dass auch musculöse Elemente sich in den bindegewebigen Verdickungsschichten der Intima der Umbilical Arterien ausbilden und eine regelmässig gestaltete Ringsmusculatur zusammensetzen."

8

moved with a small portion of the aorta and pulmonary artery still attached. Before closure the ductus is quite a tortuous canal, and, in order to get a longitudinal section which should include the whole canal, it was necessary to untwist it before imbedding the specimen for section. The sections, which showed best the relations of the ductus or ligament to the two great vessels, were those giving a profile view, that is, a section which cut the specimen longitudinally, and at the same time followed the lines of axis of the descending aorta. These sections are termed the longitudinal vertical sections. Those made in the longitudinal plane at right angles to this are called longitudinal-horizontal sections. Cross sections were also made, thus observing the structure from every point of view.

Some thirty or forty specimens were examined in detail, varying in age from the period of birth to seventy-five years. A very large number of sections were cut from each specimen, the entire ligament always being used for this purpose. In some cases, every section was stained, mounted and examined; in others, all were carefully inspected, and about fifty of the most important were selected for mounting. When cross sections were taken, the specimens were divided into several small segments, the locality of each being recorded, and in this way no difficulty was experienced in distinguishing the aortic end from the pulmonary end or the intermediate portions.

From the specimens examined, the following are selected as showing the different points of interest brought out in the investigation, the prime object being, of course, to determine the nature of the cicatrix in the aorta at the point of insertion of the ligamentum arteriosum.

DUCTUS ARTERIOSUS.

AT TERM.

Longitudinal vertical sections were made of the ductus which was still open and contained no clot.

The canal appears to be narrowest near the middle. The most striking feature of its walls is the thickness of the intima, which widens rapidly at its point of junction with the intima of the large vessels: its surface is irregular, indicating the presence of a series of transverse folds in the ductus, due, apparently, to a contraction in the direction of its axis. It consists of a homogeneous, transparent matrix, containing a large number of cells with round and oval nuclei. The cells appear to run from the deeper layers toward the surface, in rows, as if growth were taking place in that direction; there do not appear to be any cells of a distinctly muscular character in this layer.

The media also differs essentially from that of the two large vessels: it consists of bundles of cells, running chiefly longitudinally, but interlaced with circular bands of similar cells. The longitudinal bundles are seen, in the sections examined, to consist of spindle-shaped cells packed closely together, and containing elongated nuclei; the inner bundles extend the whole length of the ductus, but the outer bundles are much shorter, and this peculiarity appears to be due to the projection of the outer portion of the media of the large vessels, which overlap and enclose the outer portion of the media of the ductus. The longitudinal bundles of the ductus interlace with the elastic fibres of the media of the vessels for some little distance, and are lost in that layer. The adventitia of the vessels and ductus consist of a continuous layer of tissue. The boundary line between the intima and the media does not appear to be formed by any well-marked lamina elastica, the longitudinal bundles of cells serving to mark the inner border of the media. There are no elastic fibres in the walls of the ductus other than those which project here and there from the media of the pulmonary artery and aorta.

Remarks.—The tissue, of which the ductus is composed, is of the embryonic cellular type; there is no sharply defined outline separating its different coats, and its walls appear to be strengthened by projections from the outer coats of the aorta and pulmonary artery.

DUCTUS ARTERIOSUS.

A FEW DAYS AFTER BIRTH.

Longitudinal vertical sections were made. The openings of the ductus into the pulmonary artery and the aorta were still of the size seen before birth, but there was some contraction in the centre of the duct, which was also twisted one and one half times upon itself. Although water would still filter through, the blood had probably ceased to flow. No clot was found.

The intima thickens rapidly from its point of origin from the intima of the great vessels: in the centre it is rolled up into thick and pendulous folds. The cells, of which it is composed, are chiefly connective tissue cells, but there are undoubtedly also muscular cells to be found singly and in bundles, in this layer, such cells having been carefully observed with high powers.

Traces of the lamina elastica can be seen near the ends of the duct, but in the greater portion of its extent no lamina can be found.

The media consists of alternate longitudinal and circular bands of muscular cells, the former lying next to the intima. The elastic fibres of the mediæ of the large vessels are continued in between these bundles, where they are gradually lost, the outer fibres extending the longest distance. The longitudinal bands of the media of the ductus can also be traced into the media aortæ adjoining the lower edge of the opening into the aorta, where they disappear.

Some portions of the media and intima have begun to undergo a hyaline degeneration, but most of the tissue remains as yet unchanged.

The walls of the aorta and pulmonary artery adjacent to the openings of the ductus have as yet undergone no change. The muscular and elastic layers of the intima aortæ, which is found at the lower border of the opening, show well in this case.

DUCTUS ARTERIOSUS.

TWENTY-EIGHT DAYS.

This specimen was cut transversely. The duct has collapsed, and the lumen shows in section like two C's placed back to back; but it is still pervious. The thick intima of the ductus is still seen but has undergone a partial hyaline degeneration, a few cells, here and there, still taking the coloring matter well.

The inner surface is lined with a layer of endothelial cells. The elastic lamina shows clearly and is thrown into deep folds; but it is not always a continuous membrane. Immediately outside of this layer are bundles of longitudinal fibres, more numerous on one side than on the other; they are undergoing a hyaline degeneration. They are surrounded by a layer of spindle-shaped cells encircling the ductus, which are arranged singly and in bundles, and appear to be in a state of active proliferation. There are also a few bands of longitudinal cells in this layer. Near the pulmonary end the intima has undergone more completely the hyaline degeneration, and there are not so many bundles of circular fibres. The remains of the fœtal longitudinal fibres predominate here. There is a moderate round-cell infiltration in some parts of the media. At the aortic end the thick intima disappears, and in cross sections the arrangement of the muscular fibres appears quite irregular. The fœtal remains appear chiefly in the inner margins of the section, which shows an open lumen here; at some points fœtal tissue extends quite deeply into the middle coat.

–*Remarks.*—The ductus has collapsed but has not become obliterated. The fœtal structures are in process of elimination. A wall of new tissue is forming on the outside of the duct.

LIGAMENTUM ARTERIOSUM.

THIRTY-FIVE DAYS.

Longitudinal vertical sections were made. The central portions of the ligament consist of the walls of the ductus in a state of hyaline degeneration. A slit in the centre of this material marks the site of the former lumen. It appears to communicate with the large vessels by a narrow sinus or vessel lined with endothelium. Surrounding this degenerated tissue, and forming the outer wall of the ligament, are bands of longitudinal and circular muscular fibres, which appear to be made up both of the outer layers of the walls of the ductus, and of cells which have grown from the adjacent mediæ of the great vessels.

An examination of the aortic end shows that the opening has been closed partly by an approximation of the opposing edges of the media aortæ and partly by a growth of the intima aortæ, which is greatly thickened at this place, and fills the remaining space. Through the centre of this tissue, which is hyaline and rich in cells, runs the residual vessel above referred to. Many of the cells which are found at this point of the aorta appear to spring from the upper layers of the media, the lower layers of that coat diverging considerably to become continuous with the outer walls of the ligament. The intervening space is filled with the degenerated tissue of the ductus, which by its peculiar staining can be traced for some distance into the media of the aorta on either side.

The arrangement at the pulmonary end is not essentially different.

There is a peculiarity of the intima of the great vessels which deserves notice and exists on one side only of the openings into them; in the aorta on the lower side, and in the pulmonary artery on the upper side of the opening, or, rather, the depression marking the site of the old opening. The aortic wall overhangs the upper margin of this depression as an eyebrow overhangs the socket, and

there is a sharp curve in the fibres of the media at this spot, when
seen in section. The intima appears here as a thin layer resting
upon the elastic lamina. It becomes rapidly thickened, as it de-
scends into the depression, and helps to fill up and round out its
deeper portion. It does not, however, grow thin again, as it rises
on the opposite side of the depression and continues downward on
the lower wall of the aorta, but remains as a thick layer of hyaline
connective tissue, covered by a layer of endothelium. The line of
demarcation between the intima and the media aortæ is not distinct
on this side, as there is a layer of muscular and elastic tissue, which
appears to be placed inside of the elastic lamina and to belong to
the intima aortæ. This layer is the "musculo-elastic" layer of
Thoma, existing only, according to him, just below the orifice of
the ductus.

Remarks.—Most of the fœtal tissue of the walls of the ductus
has undergone degeneration and is surrounded by a muscular layer,
which is continuous with the media of the great vessels. The
opening into the aorta has been closed by an approximation of the adjacent walls, a thickening of the intima and a growth of cells from
the media.

LIGAMENTUM ARTERIOSUM.

FORTY-TWO DAYS.

This specimen was cut into longitudinal-horizontal sections, so
that all of the aorta seen in section was on the same level. There
is a peculiarity in this specimen not found in any of the specimens
examined. The central portion of the duct had not collapsed and
was distended with a clot. Obliteration had taken place in this
case by a closure of the two ends, and not by an hour-glass contraction of the centre of the duct, as is usually said to be the case.

The clot occupies about two-thirds the length of the ligament
and is a little nearer the aortic than the pulmonary end: it is surrounded by the walls of the ductus in a state of advanced hyaline
degeneration, no cell-structure being seen in them. This degenerative change appears to have involved the intima and the greater
portion of the media but not all of the latter coat, for its outer layer
still exists as longitudinal and circular bundles of fibres which have
been reinforced by growth from the outer portions of the mediæ of
the large vessels. At the aortic end we see the opening through the
aortic wall has become obliterated, partly by a growth from aortic

media, and partly by a growth from the aortic intima. The latter growth consists of a connective tissue composed of spindle-shaped cells in a transparent intercellular substance; it is pierced, at its centre, by a small vessel which soon breaks up into capillary branches.

The adjacent edges of the aortic media are crowded with new muscular cells for some distance and many of these cells have grown into the transparent tissue of the cicatrix. The growth of cells comes chiefly from the inner half of the aortic media, the outer half diverging to enclose the ligament and form its outer wall (Fig. 30). Between the diverging portions we can trace the hyaline degenerated fibres of the ductus.

At the pulmonary end a similar condition exists, except that the thrombus is not seen and the duct is completely closed. The approximated edges of the media and the thickened intima, pierced by a capillary vessel, show here, but the picture is not so pronounced.

Remarks.—This specimen was selected for illustration as the contrast between the different tissues was strong, but it must be remembered that the presence of a thrombus is exceptional. We see, at the aortic end, conditions closely resembling the cicatrix in an artery after ligature in continuity, namely: the slightly separated edges of the media, between which is a transparent tissue, containing spindle or muscular cells and pierced by a central arteriole.

LIGAMENTUM ARTERIOSUM.

EIGHTEEN MONTHS.

Deep dimples in the aorta and pulmonary artery mark the site of insertion of the ligament, which is bent at a right angle soon after leaving the artery, and turns to be inserted at a very oblique angle into the aorta. It is quite short. Longitudinal sections were taken in a horizontal plane.

Nearly all trace of the ductus has disappeared, a few transparent islets of tissue still, however, remaining.

The new tissue consists of bundles of spindle-shaped cells, running longitudinally, with fibrous intercellular substance. In the centre is an arteriole, or vessel of about that size, opening from the aorta, and diminishing gradually in size until it apparently opens into the pulmonary artery as a capillary vessel. This vessel at its aortic end has a distinct lamina elastica of its own, the longitudinal folds of which are quite distinct with a low power. It is lined with a delicate endothelium. The hyaline tissue through which it runs, is crowded with cells of new growth continuous with, and of the same character as, those in the ligament. The edges of the media are seen on either side of the tissue, and the muscular cell-growth is quite active, both at the edges and in the substance of the media. In the deeper layers of the intima are seen longitudinally disposed muscular cells, which run down into the cicatricial tissue. The surface of the aortic cicatrix is covered by the connective tissue layer of the intima and endothelium. At the pulmonary end the media has lost much of its arterial character and is now much less muscular in appearance.

Circular muscular fibres are still seen surrounding the ligament in its whole length.

LIGAMENTUM ARTERIOSUM.

FIVE YEARS.

Longitudinal-vertical sections were made. A sinus runs in about half the length of the ligament, apparently communicating with the aorta, but no endothelial lining could be demonstrated. At its apex a capillary vessel is given off, which communicates, in a pretty direct course, with the pulmonary artery. The upper portion of the aorta overhangs the opening, like a brow, the media taking a sharp curve backward in the section: the lower wall slants at an obtuse angle. The intima is slightly thickened at the upper margin of the orifice, and increases rapidly, in thickness, just opposite the opening, but begins to grow narrower a short distance below. It appears to consist chiefly of a connective tissue layer on which rests the endothelium. In it are seen remains of the fœtal tissue. The fibres of the media aortæ run on either side, in a shelving manner, into the compact bundle of longitudinal fibres, of which the ligament appears to be composed. Muscular cells are seen, running longitudinally, through the whole length of the ligament, but are not as numerous as in the previous specimens. A patch of calcareous matter marks the former site of unabsorbed fœtal tissue. There are bundles of circular fibres on the periphery of the ligament.

Remarks.—The process of closure is probably completed in this case, although a portion of the ligament is pierced by a sinus of considerable calibre. The fibrous character of the ligamentous tissue becomes clearly marked, and the free growth of cells into it from the mediæ of the great vessels is also noticeable.

LIGAMENTUM ARTERIOSUM.

THIRTY-EIGHT YEARS.

The peculiarity of this specimen consists in the absence of any dimple or cicatricial depression marking the point of insertion of the ligament into the aorta. The ligament is quite short and consists of a cord of longitudinally arranged elastic and non-elastic fibres. A very small vessel enters at the pulmonary end, where the usual depression exists, and extends the whole length of the ligament, terminating in a capillary network in the aortic wall. Sections were taken longitudinally and in a vertical plane, so that the aortic end is seen in profile (Fig. 31). The layers of the aortic wall, at this point, consist, first, of the usual layer of endothelium, beneath which is a thickened layer of connective tissue-cells, in transparent matrix, tapering off rapidly above and below. Beneath this lies the muscular elastic layer of the intima, so-called, the elastic elements of which are unusually pronounced, which forms over the ligament, a thick and prominent layer, and fills out the space between the edges of the media with branching elastic fibres, presenting an open meshwork, in which few cells are found. They are to be seen, however, in considerable numbers in its deeper layers, which adjoin the media. The edges of the media are still slightly separated, and taper off considerably, as they turn downwards, and become continuous with the fibres of the ligament with which the fibres of the lower aortic wall are nearly parallel. The fibres of the ligament are mingled chiefly with those of the descending portion of the aortic wall, but a portion of them take a graceful curve backwards, and are lost in the upper wall. At the point of divergence of these two bundles is a little loose tissue in which a bunch of capillary vessels is seen. At the edges of the media, which are not sharply defined, there is a growth of muscular cells, and cells of a similar character may be seen in smaller numbers between the fibres of the ligament. The outer layer of circular muscular cells is also seen in the ligament, but is now quite subordinated. A few patches of calcification appear here and there.

Remarks.—We see here the fully developed ligament, which consists of fibrous and elastic tissue, forming a dense band. At the aortic end the cicatrix consists of cells, coming from the still slightly separated edges of the media, and a thick layer of elastic fibres and plates, which fills up and obliterates the usual depression in the wall. The cicatrix in the aortic wall may be said, in this case, to

consist of fibrous and elastic tissue intermingled with muscular
cells.   It should be noted that in this case the circular muscular
fibres of the ligament are imperfectly developed.

LIGAMENTUM ARTERIOSUM.
FORTY-TWO YEARS.

The ligament is short and terminates in the characteristic depression in the vessels at each end.   Longitudinal sections in a vertical plane were made.   In the centre between the usual dense
bundles of fibrous and elastic tissue is seen a mass of loose cellular
tissue, in which a capillary network ramifies.   This tissue occupies
the axis of the ligament in its entire length, except at the points of
its insertion into the great vessels, where the longitudinal fibres are
collected into one dense bundle before they spread, to lose themselves in the coats of those vessels.   At the aortic end the upper wall
of the aorta overhangs the depression, brow-like, and, from the
apex of the depression, an arteriole makes its way to the capillary
network in the ligament. (Fig. 29.)   The elastic layer of the intima
forms a thick wall around the arteriole, but there are few cells in it.
Inside of this layer is the connective tissue layer which is slightly
thickened here: muscular cells are seen at the edges of the media
surrounding, in longitudinal and circular rows, the arteriole, and
also in the walls and between the fibres of the ligament.

At the pulmonary end there is the customary vessel, which communicates with the capillary network.   The usual condition of the
walls of this large vessel prevails in this case.

Remarks.—The appearance of the aortic cicatrix is a fair sample
of the conditions usually found.   In the centre of the ligament
the loose areolar tissue seen is found in only a limited number of
specimens.   It suggests the pre-existence of a thrombus such as was
seen in the specimen taken from a child forty-two days old.

LIGAMENTUM ARTERIOSUM.
FORTY-TWO YEARS.

Longitudinal sections were made in a horizontal plane.   The
ligament shows as a band of densely packed elastic and fibrous
tissue with few cell-elements.   In the central axis there are seen,
here and there, traces of a capillary vessel.   Its fibres divide equally
at the aortic end, and lose themselves in the central portions of the
coats of that vessel, the intima aortæ being continued over the

surface of the point of insertion, and the outer portions of the media aortæ being continued down upon the outer walls of the ligament. There is no vessel communicating with the aorta, and the depression in that vessel is a slight one. No vessel is found communicating with the pulmonary artery. There is no trace of degenerated tissue to be seen in any of the sections taken from this specimen.

Remarks.—The appearance of the aortic walls in sections taken horizontally is quite different from those taken in a vertical plane. In the former case, the media seems to form a continuous layer over the cicatrix, the continuity of its fibres being but slightly broken by those of the ligament. In the vertical section, it will be seen that there is an appreciable interval between the edges of the media.

LIGAMENTUM ARTERIOSUM.

FORTY-SEVEN YEARS.

The ligament, which was quite long, was cut into transverse sections. Sections taken as nearly as possible through the aortic cicatrix show an irregular oval-shaped mass of tissue, surrounded by the media, composed of white and elastic fibres, running obliquely towards the observer; there is no well-defined outline between this tissue and that of the media, the cells of which project into it. In the axis is a single capillary vessel, and around this very few cells are seen. Abreast of the adventitia aortæ the fibres are seen, cut transversely: the cells are more abundant: longitudinal bundles of muscular cells are seen in the centre of the ligament; and a thin band of circularly arranged cells forms its outer border. The bodies of the longitudinal cells come out quite distinctly in cross sections, as does also their division into separate bundles by trabeculæ. No elastic membrane is seen at this end.

Nearer the middle of the ligament the cells are less numerous, although the circular belt of cells is as thick as elsewhere. Islets of calcification are seen at intervals. Near the pulmonary end traces of the lamina elastica are found, and nearer still, a narrow slit, lined with a lamina, is seen. The end in section is an elongated oval, appearing to be inserted in the wall of the artery, between two layers of the fibres of the media which have been stretched apart to admit it. Cells are more numerous, again, in this end of the ligament.

Remarks.—There can be little doubt that the cells found at the ends and sides of the ligament are of a muscular character, and so probably also are many of those in the central axis.

The anatomical structure of the ductus arteriosus differs materially from that seen in any other portion of the arterial system. Some of its peculiarities are to be traced in the hypogastric artery, but exist there in a much less marked degree. At the period of birth, and before any structural change has taken place in its walls, it is a more or less tortuous canal, running obliquely downward from the pulmonary artery to the aorta, into which it opens just below the somewhat sharp curve of the lower border of the arch, at the beginning of the descending aorta. The ends of the duct are still open, but in the central portions, the walls are approximated, partly from the twisting of the vessel which is now empty, and partly from a contraction of the walls. Water will, however, readily trickle through the canal. The inner coat forms one of the most marked peculiarities of this canal, owing to its great thickness, which, however, varies considerably at different points, giving an extremely irregular wavy outline to the surface of that layer, when seen in longitudinal section. This condition is probably more apparent than real, the inequality being largely due to the twists and curves of the lumen, which render it impossible to cut this layer, in all its parts, at the same angle. The cells of which it is composed lie in a transparent intercellular substance, and are spindle-shaped: they are arranged for the most part longitudinally, although this is not always apparent, owing to the irregularities mentioned. By some observers they are supposed to be connective tissue cells, by others, muscular. The more superficial cells do not, as a rule present the type of the muscular cell, but very perfect examples of the muscular cells are seen in the deeper layers of the intima. The boundary lines of the different coats are exceedingly indistinct. The lamina elastica is not easily found in the longitudinal sections, and does not appear to be continuous, being wanting at many points. It is more easily made out in cross sections.

The media consists chiefly of bands of longitudinally arranged muscular cells; these are occasionally separated from one another by circular bands of muscular cells chiefly at the outer border of the vessel. The coat is almost entirely a cellular one. The few elastic fibres which it contains can be traced into the aorta and pulmonary artery; they are most abundant in the outer layers of this coat, and occasionally extend throughout its whole length.

At the aortic opening the intima narrows rapidly on either side; at its upper margin the aortic wall makes a sharp curve upward, when seen in longitudinal-vertical sections, to form the lower curve of the aortic arch. Here a thin layer of cells lying upon the elastic lamina, which is well defined, forms the intima aortæ. The walls of the aorta, at the lower margin of the opening, are thicker, and are continuous in a nearly straight line with the tissue of the ductus; the tissues of the coats of the aorta and ductus are interwoven at this point, and those of the ductus are spread out in a fan-like shape, and are lost in the different layers of the wall of the aorta. The elastic lamina of the aorta does not form here a continuous layer, as above, but is broken into several more or less parallel layers. According to Thoma, that portion of the wall which lies inside of the principal elastic layer, belongs to the intima, and is described by him as the musculo-elastic layer of the intima which forms a reinforcement to the wall at this particular point. A similar arrangement of the elastic lamina is found elsewhere in the aorta, and it would seem to have no significance at this spot further than as an indication of the tendency of the elastic tissue to form less of a limiting membrane in the neighborhood of the ductus, where all layers are ill defined. The intima of the ductus narrows rapidly, as it emerges at the lower border, and becomes continuous with a similar layer of cells, here appearing as connective tissue cells, which form a narrow layer in the intima of the descending aorta. This layer was traced about half an inch below the opening of the ductus, at various periods from birth up to adult life, but was not found to form the thick layer described by Thoma as lining the interior of the aorta between this point and the iliac artery after birth.

The arrangement of the walls of the pulmonary artery resembles that seen at the aortic opening, but in a reverse order. Here, the upper wall is the thicker, and it receives the tissues of the upper wall of the ductus which are freely interlaced with it. The difference between the two margins of the pulmonary opening is not so marked as at the aortic end.

A few weeks after birth a very marked change has taken place in all the tissues of the ductus, which appear to be now undergoing hyaline degeneration, preparatory to absorption. The outermost walls alone remain, but it is not certain whether the structures, still preserved from degeneration, should not be regarded as belonging to the aorta and pulmonary artery. A few weeks later, this outer wall is found to have become greatly strengthened, and now

forms a layer of circular muscular fibres, which enclose the tissues
of the ligament, and is continuous with the mediæ of the two great
vessels.

At this period the margins of the media aortæ at the aortic
opening are greatly approximated by the obliteration of that orifice.
The intima forms a thickened layer which partially fills out the
umbilicated depression in the aortic wall and the space still intervening between the ends of the media. (Fig. 30.) Here it forms a
hyaline tissue pierced by a small arteriole apparently of new formation, its cellular elements, consisting of cells growing from the intima
and media. The latter coat is divided into an inner and an outer
layer which diverge slightly, the inner directed toward the cicatricial tissue, the outer being continuous with the outer layers of the
ligament. Between these two layers lies the degenerated tissue of
the ductus (Fig. 30). From the inner layer an active cell-proliferation is taking place, and with high powers a beautiful mass of
muscular cells is seen, growing in profusion, leaving little doubt as
to the part played by the media in the healing process.

In Fig. 30, the walls of the ductus are represented as open in
the middle portion; in the specimen the space was occupied by a
thrombus. The formation of a thrombus does not appear to be
usual, and is probably due to a more active contraction at the ends
of the ductus, than at its centre. In many adult specimens a cleft or
sac is seen occupying a considerable portion of the centre of the
ligament, the presence of which is probably due to the pre-existence
of a thrombus.

The specimen taken from a child eighteen months old shows a
complete development of the ligamentum arteriosum, traces only
of the hyaline degenerated tissue of the ductus remaining. Later
in life, the tissues do not appear to undergo any essential change
other than a degenerative one, calcification of certain portions of
the ligament being seen at all ages and being probably due to further
change in islets of unabsorbed fœtal tissues.

When fully formed, the ligament consists of a dense bundle of
longitudinal fibres composed of fibrous and elastic tissue; these
contain a moderate number of cells of spindle shape; this tissue
is enclosed in a layer of circular muscular fibres of varying thickness. Outside of all is the adventitia reflected from the coats of
the great vessels. In the central axis of the ligament a small
vessel is usually found which can be traced, either directly or
through a few capillaries, to the aorta or pulmonary artery; occa-

sionally a cleft or blood space, lined with endothelium, is seen connecting at either end with a small vessel. Less frequently the bundle
of fibres of which the ligament is composed is divided, a mass of
loose areolar tissue occupying the space between them; this tissue
usually contains a rich capillary network. The areolar tissue, like
the sinus, occupies only the middle portions of the ligament, which
is closed with a denser tissue at either end. At the aortic end of
the ligament, when studied in longitudinal vertical sections (Figs.
29 and 31,) the ends of the media aortæ are seen still slightly
separated. The fibres of the ligament seem to be more directly
continuous with the lower wall of the aorta. In the greater number
of specimens the indentation of the wall which marks the site of
the opening remains permanently, and can be seen as a dimple situated beneath a sharp ridge or brow, formed by the inferior wall of
the aortic arch, just beyond the point of origin of the occipital
vessels. The edges of the media aortæ on either side are turned
slightly outwards to meet those fibres of the ligament which lose
themselves chiefly in the lower wall. Here the fibres can be traced,
not only into the media, but also into the musculo-elastic layer,
so-called, of the intima; some of the fibres curl upward to be inserted into the wall at the upper margin of the opening, and
the space between the diverging bundles is filled out by a growth
from the more superficial layers of the intima. It is here that the
small arteriole branching from the aorta is usually found. The
cells are much more numerous in the tissues just described than
elsewhere in the ligament, and many of them are of a distinctly
muscular character; bundles of them being traced into the media
and musculo-elastic layer Many of them can be seen surrounding
the small central arteriole.

Occasionally no aortic depression is found, the surface being
perfectly smooth, so that no indication is given of the exact site of
the aortic cicatrix. In these cases an unusual development of
elastic tissue appears to have taken place from the musculo-elastic
layer, so that the depression between the everted edges of the aorta
has been completely filled up (Fig. 31). In some cases, all direct
vascular communication appears to have ceased.

At the pulmonary artery end there is, invariably, a symmetrical
depression, and the nature of the healing of that wall does not
differ essentially from the most common form of aortic cicatrix.

An analysis of the nature of the cicatrix in the aortic wall, the
feature of the ligamentum arteriosum of most interest, in the pres-

ent connection, gives a combination of tissues closely resembling that found in the cicatrix of large arterial trunks after ligature in continuity. The slightly separated edges of the media stand out distinctly from the other tissues: between them lies the new tissue, which, in some cases, consists of a hyaline basis filled with muscular cells, and perforated by a central arteriole. This layer tapers off, on either side, becoming continuous with the intima; at other times, the hyaline tissue is partly replaced by a considerable formation of elastic tissue, a condition not seen in arteries, but which one would expect to find in a vessel containing so large a predominance of elastic fibres as the aorta. The presence of a layer of circular muscular fibres in the ligament, does not seem to bear out the analogy, which might be expected, with the cord which unites the two ends of an artery. In the largest, and most powerful arterial cicatrices, we find the new tissue enclosed for a considerable distance by an arterial wall (Fig. 27), giving the support needed at a point where the blood pressure would cause a severe strain upon the cicatricial tissue. In the aorta, the obliquity of the insertion of the ligamentum arteriosum, and the overhanging edge at the upper margin of the cicatrix, give a valve-like action to the ends of the media aortæ which are pressed together, rather than separated, during the dilatation of that vessel.

A study of the obliteration of the ductus affords additional proof of the power of the muscular elements to form new tissue. The activity of these cells was observed in a number of specimens, examined a few weeks after birth, and the developed cicatrix shows bundles of cells, which not only resemble muscular cells, in their shape and size, but which can be traced directly into the media, and the musculo-elastic layer of the intima, where muscular cells are always found in abundance. The changes seen in the wall of the ligament are alone sufficient proof of the power of this cell to take part in formative processes, a rôle which, curiously enough, has been almost universally denied to it by pathologists.

THE HYPOGASTRIC ARTERY.

The other vessel closed by nature at the time of birth is obliterated under circumstances which correspond pretty accurately to those which obtain in the artery of an amputation stump. A large portion of the territory supplied by the vessel is permanently removed, and the blood which now flows through it is distributed to an area of comparatively limited extent. There is no longer need of a vessel of the calibre of the hypogastric artery, and the changes, which ensue, should affect the vessel in its whole extent. That such a result follows is a well-known anatomical fact; the remains of the hypogastric artery subsequently doing service as the superior vesical artery. The observations of Baumgarten are intended to show that the old vessel remains entire, and is not destroyed as had been taught by Virchow. Thoma's studies have led him to express the opinion that a growth takes place from the intima, by which the calibre of the vessel is narrowed, and that new walls are thus formed within the old.

The following examples are selected from a number of observations, made to determine the precise nature of these changes, and to supply a standard with which to compare the investigations made upon the healing of arteries after amputation.

On laying open the abdominal cavity of an infant, the hypogastric artery can be readily traced from the umbilicus to the brim of the pelvis, just beyond which point it merges, with the internal iliac artery, into the common iliac. The upper portion of the internal iliac artery, an exceedingly small vessel at this period, is for the time being, appropriated by the placental circulation. In removing the specimen, the point of junction of the hypogastric with the internal iliac was carefully sought for, and the specimen preserved from this point to the umbilicus for histological study. Each specimen was divided into segments, so that the different portions of the vessel could be studied separately. Both longitudinal and transverse sections were made. The examples presented illustrate the earliest changes which take place at and soon after birth, as well as the completed process in the adult. The specimen taken from the monkey is reported, both for the purpose of a comparative study, and on account of the perfection of its histological details.

THE HYPOGASTRIC ARTERY.

INFANT A FEW DAYS OLD.

On laying open one of the pair of vessels, a clot about one and one quarter inch in length was found at the umbilical end of the vessel. The other vessel was removed from the point of bifurcation of the common iliacs, and cut into four equal portions. The first portion extended from the bifurcation of the common iliac to the junction of the hypogastric artery with the internal iliac, and was cut into longitudinal sections.

9

The only change to be seen here is an appearance of slight atrophy of the portions of the media, that coat taking the coloring matter less perfectly than other parts of the vessel. The difference between the lamina elastica of the internal iliac and its equivalent in the hypogastric artery is striking. In the former it constitutes a thick and glistening membrane; in the latter, there are several elastic layers close together, no one of which is well pronounced, and between them are longitudinal rows of muscular cells.

In the second portion, consisting of the next adjoining segment, also cut into longitudinal sections, we trace the elastic layers just mentioned, but find no well-defined lamina elastica. There is as yet no change in the intima.

In the third portion, beginning at the middle of the vessel and extending towards the peripheral end, cut transversely, we meet with numerous bundles of longitudinal, muscular fibres in the media. The vessel is now greatly contracted, and the wall is thrown into longitudinal folds. Near the point where the clot terminates there is a distinct thickening of the intima, apparently due to the proliferation of the endothelial cells. These cells are firmly attached to the apex of the thrombus. There is, as yet, no outgrowth into the clot or lumen from this layer.

In the fourth portion, or that part attached to the granulating surface of the umbilicus (cut longitudinally), we find the terminal portion of the vessel, for about one quarter of an inch, very tortuous, collapsed, containing no clot, the end lying exposed among the granulations. The walls are in a state of incipient hyaline degeneration; that is, the elements of the tissue take the coloring matter very feebly, their outline is imperfect, and there is a general condition of transparency pervading the tissues of the vessel. This condition extends to a point about three quarters of an inch from the surface; beyond this we see the cell structure distinctly, as before. This part of the vessel differs from the more proximal portions in the diminished amount of elastic tissue, the vagueness of outline between the inner and middle coats, and the amount of muscular fibre which is somewhat greater here.

Remarks.—We have here an example of the earliest changes to be seen after the circulation in the umbilical cord has ceased. A thrombus fills almost one quarter of the lumen of the vessel, a slight growth of endothelium, serves to attach the clot firmly to the vessel-wall, and the extreme terminal portion of the vessel is undergoing a hyaline degeneration.

HYPOGASTRIC ARTERY.

INFANT THIRTY-SIX DAYS OLD.

The umbilicus was removed together with the iliac arteries. At their terminal portions the two hypogastric vessels united to form a cord of hyaline tissue attached to the umbilicus. The lumen had become obliterated near the point of junction. An inch of each vessel was removed at this end, and sections were made transversely in one, and longitudinally in the other.

Longitudinal sections include the closed end of the vessel. The lumen is tortuous, making it difficult to obtain a cut of a continuous portion of the vessel of any length. The outer coats of the vessel are, however, quite straight. These consist of embryonic connective tissue, of which the cord, to which the two vessels are attached, is also composed.

There are both longitudinal and circular fibres in the media. These close around the stump of the vessel, and become interlaced. There are very few elastic fibres, and the appearance of an elastic lamina is seen only at one or two points.

The clot is penetrated by columnar masses of embryonic tissue which spring from the inner walls: the intima appears as a layer of hyaline connective tissue containing spindle-shaped cells, and is separated from the media by a very thin and interrupted elastic membrane.

Cross sections bring out the circular fibres: in those near the distal end, the cells are undergoing a hyaline degeneration which pervades the whole wall. The lumen is much reduced in size, and has a stellate outline. The line of separation between the inner coats is ill defined low down, but higher up the elastic membrane is found, as are also elastic fibres between the bundles of cells. Granulations are seen penetrating the clot, and deep folds of the walls also project into it, each helping to obliterate the lumen.

Near the origin of the vessel cross sections show a contraction of the lumen, throwing the walls into irregular folds, which are bridged over by a slight growth of young connective tissue on their inner surface. Where a branch is given off, the typical lamina of the latter may be traced directly into the tortuous and interrupted elastic membrane of the hypogastric artery.

Remarks.—The ends of the vessel, for an inch or more from the umbilicus, show signs of a hyaline degeneration. One quarter of this portion has already been converted into a cord. The lumen is greatly narrowed for a considerable distance beyond by a contraction of the media, and by a growth from the walls of the vessel into the thrombus, which is in process of disintegration. Near the origin we see the beginning of a growth which is destined to diminish the calibre of the vessel at this point.

REMAINS OF THE HYPOGASTRIC ARTERY.

Adult.

The specimen included a portion of the internal iliac, the superior vesical artery, and the ligament to within about three inches of the umbilicus.

One of the vessels was laid open from its origin in the internal iliac for about one third of its length, when the lumen became too narrow for the microscope-scissors to follow it farther. The walls were found to be much thicker than in vessels of this size elsewhere. The lumen could be traced by the naked eye for some distance farther by cutting open the specimen with a sharp razor. Beyond this point, the cord appeared to consist of bundles of fibres surrounding a centre of denser and whiter material, which could be traced to the end of the specimen; that is, to within about three inches of the umbilicus. The companion vessel was divided into nine portions for microscopical examination, cross sections being made of each portion, except the first, which included the point of junction of the superior vesical with the iliac artery, and was cut longitudinally.

At the point of origin of the vesical artery there is a thickening of the intima of the internal iliac, which could be traced to the end of the specimen, about one quarter of an inch, in the direction of the origin of the iliac. This thickening, which consists of circularly and longitudinally disposed spindle-shaped cells in a hyaline matrix, appears to originate between two layers of the elastic lamina. The internal lamina accompanies the new tissue into the vesical artery to its most distant portions, and here, at all events, appears as a new formed membrane; a membrane which is not found in foetal life. It is at first thrown into deep transverse folds, and becomes more widely separated from the middle coat in the vesical artery.

Cross sections in the second portion of the specimen show the following layers in the walls of the vessel (Fig. 32). On the inner surface is the delicate layer of the endothelium, beneath which is the elastic lamina, a thick but not highly refracting membrane, such as is usually seen in other vessels of this size. It is slightly undulating in character, and, in this respect, is in contrast with the deep folds into which that membrane is usually thrown in contracted vessels. Outside of this lamina is a narrow layer of circular muscular fibres, on the outer borders of which are found here and there a few longitudinal cells. Next comes the imperfectly developed lamina belonging originally to the hypogastric artery. This is thinner and not a continuous membrane, but refracts the light more powerfully than the inner membrane, and is thrown into deep folds. Outside of this lamina is the old media of the hypogastric artery, consisting of cells disposed circularly, obliquely, and longitudinally. This coat varies greatly in thickness at different points, and frequently on different sides of the vessel at the same point. Outside of all lies the adventitia. These various layers make up a wall thick out of all proportion to the lumen of the vessel.

In the third portion examined the conditions described remain unchanged; but in the fourth portion the lumen grows rapidly narrower, the outline between the double walls becoming indistinct and difficult to trace. The course of the vessel is now tortuous, and soon loses itself in the tissues of the ligament. A more careful examination of sections at this point, however, shows a slight cleft in the fibres, of irregular outline, surrounded by a denser tissue than prevails in other portions of the section, containing remains of the elastic fibres and membrane. These remains of the vessel can be traced through the fifth, sixth, and seventh portions, becoming gradually fainter, and are with difficulty found in many sections.

In the eighth and part of the ninth portion, where the cord becomes more compact, and the tissue takes coloring matter better, it is more distinctly seen again, and, in the central cleft, a capillary vessel can be traced for a short distance: finally, in the most distal sections it is not possible to say whether any traces of the vessel can be found.

Remarks.—Traces of the old vessel can be followed to within three inches of the umbilicus. We have evidence of contraction of the outer walls, followed by a development of new tissue inside the vessel. In this respect, and in its gradual diminution in size to an arteriole and capillary, it closely resembles the artery of an amputation stump.

HYPOGASTRIC ARTERY.

MONKEY.

Both vessels were removed, but the vessel of one side only was examined: it was cut into four portions, cross sections being made of each portion. Near the origin of the artery a growth of new tissue is found on the inner surface of its walls, between the endothelium and the elastic lamina. It consists of spindle-shaped cells in two layers, the inner arranged circularly; the outer running parallel to the axis of the vessel. This growth gradually increases in quantity as we approach the distal end of the vessel; and, on arriving at about the middle portion, we find the lumen diminished to one half of its calibre by the new growth; the remaining lumen seeming to owe its existence to the presence of a newly formed vessel occupying the centre of the new tissue, (Fig. 34), and possessing walls of its own independent of those which pre-existed.

We find here a delicate endothelium surrounded by a layer of muscular fibre, no elastic lamina intervening between the two layers. Around this central vessel is what may be regarded as the obliterating growth, composed of connective tissue, and, possibly, some longitudinal muscular cells.

The old elastic lamina, which is a very thick and glistening membrane, seen throughout the whole course of the vessel, comes out in strong relief outside of the new tissue, but is generally found ruptured at one or two points. It is thrown into unusually deep folds. Outside of it the cells of the media are seen, arranged circularly, and differing from the media in man, where longitudinal fibres are also found. There is nothing worthy of mention in the appearance of the adventitia.

Near the terminal portions the vessel appears much contracted, becomes tortuous, and the various layers are less easily distinguished. Finally, we come to sections where arterioles and capillaries only are found.

Remarks.—The changes observed in the interior of this vessel closely resemble those seen in obliterating endarteritis. The perfection of the staining process, and the clearness of the cell-structure, show the muscular character of a portion of the obliterating tissue. No newly formed elastic lamina is seen in this specimen. The vessel is, however, very much smaller than that taken from the human subject.

On laying open the hypogastric artery of the newly born child, a thrombus is found, extending from the point of obliteration at the umbilicus through about one third of its extent. This was the case in all specimens examined. Considering the very moderate amount of traumatism exerted upon this vessel, the presence of a thrombus is of special interest in its bearing upon the question of the formation of a thrombus. It is not probable that the clot takes its origin in the extra-umbilical portion of the vessel after the ligature has been applied, for this portion contracts rapidly, and its lumen is obliterated very shortly after birth. The very moderate amount of inflammation produced at the umbilicus could hardly suffice to produce a clot of such size; for, as a rule, healing takes place without the formation of pus. So far, therefore, as these observations go, no very large number of specimens having been examined, they fail to support the theory of the traumatic origin of the thrombus.

The hypogastric artery presents certain peculiarities which distinguish it from other arteries. The most striking of these is the presence of a large amount of longitudinal muscular fibre, in which respect it bears a resemblance to the structure of the ductus. It has another peculiarity, also, in common with that vessel, consisting in the absence of a well-defined outline to the inner wall of the media, which a well-formed lamina elastica gives. Elastic tissue is found at this point, but the membranes are thin, not always continuous, and sometimes hard to find. Near the distal end of the vessel there is little to be seen of any such structure.

When the placental circulation ceases, a marked contraction takes place throughout the greater part of the vessel, and its most distal portion is filled with a thrombus. While the healing of the umbilical cicatrix is taking place, the distal end of the artery is undergoing hyaline degeneration which pervades its whole thickness. A few weeks after birth we see that portion where the two vessels lie in contact, reduced to a cord of gelatinous tissue in which all traces of the vessel have disappeared. This change extends, and, at this period, has already involved the walls of each artery for perhaps one half to three quarters of an inch beyond this point, indicating evidently a destruction of the vessel, for a considerable distance from its distal end. This view is borne out by the examination of the fibrous cord, which attaches the superior vesical artery to the umbilicus. Here, traces of the walls of the vessels are still seen, chiefly fragments of elastic, and dense, fibrous tissue. Baumgarten's view, that the vessel is preserved entire, but appears smaller owing to the growth of the body, is thus disproved, and the possibility of the destruction of a considerable fragment of the hypogastric artery is clearly established.

In the early days of life but little change is seen in the interior of the vessel; a slight proliferation of the cells of the intima is noticed near the apex of the thrombus, but not elsewhere. By the second month, however, a distinct growth of tissue may be observed throughout the entire length of the inner wall. The coats have contracted, and the inner surface is thrown into deep folds which are bridged over by a growth of young cells, lying imbedded in a hyaline, intercelullar substance. This tissue fills out the irregularities, and, in cross section, the lumen now presents a smooth contour. In the region occupied by the thrombus the growth of new tissue is most active; granulation-like masses intersect the clot, and are rapidly obliterating the lumen. At this period we are able to establish the fact of a general contraction of the vessel throughout its whole extent, a still further narrowing of the lumen in its proximal portion, and the beginning of a destructive process at its terminal portion.

In adult life, a cross section taken through the walls near the proximal end, shows the old wall of the vessel much contracted, as indicated by the deep folds of the elastic lamina (Fig. 32). Within exists a tissue evidently formed subsequently to this contraction, for it does not appear to have been affected by it. This tissue has now developed into a newly formed lamina elastica, surrounded by a new layer of muscular cells. Although the lumen is quite small, the walls of the vessels are of considerable thickness. Tracing the vessel to its extremity, we find the lumen constantly diminishing in size, the newly formed internal coat becoming less distinct, until the structures examined consist of a cord of fibrous tissue in which a tortuous arteriole finally breaks up into smaller vessels which are not to be distinguished from capillaries. Beyond this point we see only a fibrous cord, in the centre of which is some denser, more opaque tissue intermingled with traces of an elastic membrane.

The series of changes which has taken place since birth in this vessel may be summarized as follows: a contraction followed by a still greater diminution of the calibre by an obliterating growth of the proximal portion, and a complete destruction of the terminal portion, in extent amounting to about one third of the original vessel. Thoma's assumption that a portion of the newly formed tissue is of a muscular character is fully confirmed. That this tissue is, however, developed from the intima, as he asserts, can hardly be accepted, the imperfect separation of the intima from the media making it quite possible for that layer to have participated in the growth. This view is in a measure confirmed by the investiga-

tions quoted above, bearing upon this point. Baumgarten's statement, that the extreme end of the vessel is closed with fibrous tissue
is not borne out, the vessel gradually diminishing in size to an
arteriole which, in its turn, ends in capillaries.

Cross sections of the artery taken from the monkey show the
newly formed muscular layer with great distinctness. A new lamina
does not exist, but the old wall contains a strong lamina thrown into
deep folds (Fig. 34). In this respect, it corresponds with the arteriole seen in the obliterating tissue of the tibial artery (Fig. 35),
where the lamina is also preserved in the old vessel-wall, but absent in
the new wall, differing, however, from the arrangement in the hypogastric, in which the old lamina is not a well-developed membrane.

It is also worth noting that the walls, both of the obliterated
tibial artery and femoral, have undergone calcification (Figs. 35 and
33). Such a change was not observed in any of the hypogastric
arteries of adults which were examined, but was almost invariably
found in the ligamentum arteriosum. The observation suggests the
view that this degenerative change may be the result of cessation of
function of the vessel-wall, rather than the sign of disease which
has led to obliteration of the vessel. If this be the true explanation
of changes in the walls of the vessels, the subjects of endarteritis
obliterans, new light is thrown upon the etiology of that disease,
which should no longer be regarded as originating in an inflammation of the coats of the vessel. Such changes as are observed are
due more probably to a participation of these vessels in a formative
process made necessary in consequence of diminished blood supply,
and their subsequent degeneration from lack of use. A study of
the changes observed in the hypogastric artery after birth brings
out a close resemblance to the changes seen in arteries after amputation. In the earliest period we find contraction of the vessels in
their entire extent, with occlusion of the distal portions by thrombosis.
Later, a growth from the inner wall is observed in each case, and
finally new blood-channels are established within the old, the newly
formed vessels being accurately adapted to the decreased blood
supply. In the hypogastric artery there is the most satisfactory proof
of the formation of new muscular fibre, which, taken with the observations in amputation stumps, shows that, even in this form of
healing after ligature, new muscular fibre may be produced.

CHAPTER V.

SUMMARY.

THE materials collected by these experiments and investigations are, we think, sufficient to enable us to draw a few conclusions as to the nature of the process of repair in arteries after ligature, and to recognize some of its more important phases. One of the most interesting facts thus established is the duration of the process, which greatly exceeds that usually ascribed to it. The period of time from the moment of ligature to the development of the final scar, varies considerably with different vessels. For those of large size it may be said to range from three to six months. In some cases a cicatrix may be formed at an earlier date, but in no instance has it been possible to find a specimen in which all the changes of the series have been fully completed before the shorter period.

Another feature brought out quite prominently is the complicated nature of the process. The walls of a vessel of the first class are composed of structures of great strength and density, the elastic tissue, of which they are largely constituted, being one of the strongest and toughest tissues in the body. The vasa vasorum are minute vessels, and ramify in a tissue which is, within certain limits, quite unyielding to pathological influences. The conditions favorable for a rapid series of changes dependent upon inflammatory and reparative processes, such as is found in the more loosely woven connective tissues, do not exist here, and nature is consequently obliged to resort to the expedient with which we are most familiar in connection with the process of repair in bone, of supplying provisional structures which seal the vessel while the coats gradually elaborate those elements which are to form part of the future cicatrix.

Many of the specimens preserved in museums have the appearance of having united by first intention. Those illustrating the precise moment when the adventitia has been absorbed by granulation tissue, and the ligature lies imbedded in the tissue, separated from the two ends of the vessel (Fig. I, Frontispiece), suggest strongly such a mode of union. Some writers have advocated the possibility of such an occurrence, but there are no facts in the materials before us which authorize any such conclusions.

Finally, the nature of the cicatrix observed in the foregoing specimens differs essentially from that usually ascribed to it. No question in surgical pathology has been more vigorously debated than that which refers to the part played by different elements in the process of repair in arteries. On one point there seems to have existed a singular unanimity, it being universally conceded that the muscular coat did not participate in the process. The experiments here recorded seem to show pretty conclusively that the muscular cell is a prominent and essential feature of the arterial cicatrix.

The process of repair may, for the purpose of study, be divided into three principal stages. In the first are included those changes which occur immediately after the application of the ligature, and during the period in which the walls of the vessel are gradually separated from it. The second stage embraces the period of development of the provisional structures, or, if we prefer to use the term, the external and internal callus. During the third stage there is a process of differentiation of the elements of which the internal callus is composed, and a gradual absorption of the provisional tissues until the period of final cicatrization is reached.

The application of a strong ligature to the trunk of an artery usually ruptures the inner and middle coats; this is always the case in dogs, and the human specimens examined showed no essential difference. The outer coat is composed of elastic tissue which is collected into a dense and singularly strong bundle of fibres, which protects the injured walls and prevents hemorrhage. Some idea of the powers of resistance of this fascia-like tissue may be gathered from the experiments of Ogston. The ruptured inner walls are slightly inverted so as to form a V-shaped wound when seen in section (Figs. 1 and 2), or a cone with its base presenting towards the lumen of the vessel. There is no valve-like incurvation of the walls, as is said to exist after torsion. The proximal end, if the thrombus be large, assumes an ampulla-like shape, as described by Bryant. This is probably due, not to a dilatation of the vessel, but rather to an inability of the vessel to contract, owing to the size of the thrombus, which is largest at its base. The walls, however, are stretched and thin, compared with other points, indicating that the vessel here is of full size. The distal extremity is more uniform in calibre, and contracted as much at its end as elsewhere, or nearly so.

The size and appearance of the thrombus differs greatly in different specimens. In ligatures in continuity in man, of the period during which the thrombus is found, it was a marked feature of the preparation. It was both "red" and "stratified" in appearance. In all cases it extended up to the first branch of considerable size. Usually it was firmly adherent at its base to the walls of the vessel on the proximal side, and, as a rule, the distal thrombus became separated from the specimen in course of preparation. The proximal thrombus was not only wider, but also longer than the distal thrombus in all these specimens. In the amputation stumps the vessels, as far as removed, were in most cases filled with thrombi. All the specimens examined were taken from septic wounds.

In dogs there was a great divergence in size. In some cases the entire trunk was filled with clot, in others the thrombus was less than two millimetres in length. It was invariably a "red" thrombus, but was not always stratified. In the horse the amount of thrombus also varied greatly. In the specimen shown in the Frontispiece the proximal thrombus is a white or "decolorized" thrombus, the two thrombi being about equal in size.

The series of ligatures made under antiseptic precautions produced thrombi of greatly diminished size, the proximal thrombus being about one and one-half millimetres in length in the smaller specimens observed, the distal being slightly smaller. Such specimens were taken from vessels about three and one-half millimetres in internal circumference. In no case was the thrombus absent. When the thrombus was of the size just mentioned it was polypoid in shape, and was attached by its pedicle to one of the projecting edges of the media.

The marked contrast between thrombi in wounds where antiseptic precautions were observed and union by first intention was obtained, and those in which no precautions were taken, was such as to leave no doubt as to the influence of septic products upon the process of coagulation. In all cases the amount of bruising of the vessel-wall was about the same as has been described. In certain attempts to close the vessel without rupturing the internal wall, it was found that the lumen had not been fully obstructed, and, at the same time, that the walls had been bruised to a certain extent, so that, whenever the ligature was drawn tightly enough to obstruct the flow of blood, there was considerable bruising of the media. Although in two cases examined, the opening through the ligature had been extremely small, and although there was considerable, but not continuous, bruising of the media around the vessel, no trace of thrombus was observed in either case. The experiments were made on the carotids of large dogs, where the current was strong.

The experiments of Baumgarten, Senn, and others tend to show that the extent of traumatism is the most important factor in the production of the thrombus, the exciting cause being probably the presence of some ferment-yielding substances. The experiments with antiseptic precautions, reported above, are confirmatory of such views. In some of the preparations, taken from extremely septic amputation wounds, the thrombus was exceedingly small; on the other hand, a clot of considerable size was always found in the hypo-gastric artery of infants newly born, filling nearly a third of the vessel, and, in one case, a thrombus of considerable thickness was found to fill about two-thirds of the obliterated ductus arteriosus. Certainly in the latter case there could not have existed a local sepsis or traumatism. In none of the experiments performed was there an entire absence of the thrombus. In one or two experiments, made with the double ligature, the blood did not remain fluid between the ligatures, as has been reported by a number of observers, although the wounds healed by first intention. Baumgarten maintains that an artery can be tied without the formation of a thrombus. In some of the experiments, recorded above, the thrombus was about the size of a mustard-seed, and could easily have been overlooked, or might accidentally have become detached during the preparation of the specimen. We must conclude, therefore, that although traumatism or sepsis may be the chief factors in the production of the thrombus after ligature, the size of the thrombus does not depend entirely upon the extent or degree of these conditions. Granular masses or " blood-plaques " were frequently observed in experiments upon dogs. (Figs 6, 7, 10.)

The rapidity of the formation of the thrombus was a question considered during experimentation. In the femoral artery of the dog, ligatured in continuity, there was no thrombus one hour after the ligature had been applied. In the brachial artery of man, which had been lacerated, there was a firm homogeneous clot two hours after the injury had been received (Fig 4). In the femoral of a man, whose thigh had been amputated twenty hours before, there was already a well-formed laminated clot. In vessels in which the hemorrhage has ceased spontaneously, a thrombus will probably always be found immediately after the bleeding has stopped. After ligature it is probable that an hour or more may elapse before the thrombus has become perceptible to the naked eye. That a human artery can be tied and heal without the formation of a thrombus, is a possibility which the data at present in our possession do not authorize us to concede. The thrombus protects the wound in the vessel. This is evident when the inflammation is severe and suppuration takes place. Its period of greatest usefulness begins when the walls of the vessel are retracting, and granulations are growing into it. The thrombus then serves as a good culture medium for the granulation cells in its deeper portions, while its extremity forms a protecting scab.

Soon after the application of the ligature a collection of wandering cells takes place in its neighborhood, and it is presently imbedded in a mass of granulation tissue which surrounds the ends of the vessel. At an early day some of these cells may be seen in the meshes of the adventitia, where the fibres are not too closely packed together, and in the outer layers of the media. At the base of the proximal thrombus, which is firmly attached to the wall of the vessel, are also numbers of white blood corpuscles, which are most numerous in and about the little V-shaped cleft in the wall formed by the rupture at the point of ligature. The appearances are suggestive of an immigration of wandering cells, but usually these cells have only penetrated the outer walls of the vessel, and the collection of cells seen within the vessel is probably due to the accumulation of white corpuscles at the bruised spot during the process of coagulation. Occasionally (Fig. 1) we may see, however, an invasion of the walls by wandering cells from without, and it is probable that a limited number of wandering cells find their way into the interior of the vessel in every case, during the first week.

In the interior of the vessel we find no marked activity of the cell-elements during this period. In many specimens there appears to be absolutely no change whatever in the endothelium. In others, and probably in the majority of cases, a proliferation of these cells may be observed, particularly at the distal end (Fig. 3). A somewhat more active growth may be found at the apex of the thrombus, where the vessel-walls have contracted closely around it. Here a distinct thickening of the intima may be seen, not only in dogs but also in man, and a number of spindle-shaped cells can be found attached to the surface of the clot. Wherever, at some point more or less remote from the ligature, ruptures have occurred in the lamina elastica, and these are common, a growth of cells can be observed projecting from the media for a short distance into the clot (Fig. 6). The total amount of cell-growth produced from these various sources is insignificant, and its chief purpose seems to be to attach the thrombus more firmly to the inner walls of the vessel, so that, when the walls escape from the ligature, and expansion takes place, its connection with the vessel-wall shall not be seriously affected. A more prominent place in the process of repair cannot be given either to the endothelium or to the wandering cells. Later, a growth from the endothelium probably supplies a coating to the new vascular spaces in the cicatrix, but an extended series of observation fails to show that these cells play the principle rôle in the formation of the cicatrix, as is maintained at the present time by a large number of histologists.

The special experiments with double ligature, following the method of previous observers, in order to test the activity of the wandering cells or of the endothelium, failed to show that either of these groups of cells were the agents by which the segments became obliterated, but showed that granulations eventually grew in at each end, and filled this portion of the vessel. A slight proliferation of the endothelium was occasionally observed, but in other cases it was absent entirely, and in no case did the cell-growth amount to a perceptible thickening of the intima. In severe forms of inflammation it was found possible for a wandering cell to enter the segment cut off by the ligatures before the ends had admitted granulations.

In some specimens a slight growth of endothelium was seen covering the bruise in the media, when the clot was small and permitted a view of the part. In the temporary ligature, the wound in the media was found, at the end of the first week, covered by a thin film of endothelium, and active cell-proliferation was found also in the media.

In none of these various experiments was there any evidence to show that the cell-growth in the interior of vessels during cicatrization was supplied from the intima alone.

Outside the vessel the granulation tissue accumulates to such an extent as to cover in the ligature and the ends of the vessel, forming a spindle-shaped mass, thickest over the ligature, and tapering off above and below. It varies greatly in size according to the amount of inflammation produced. In healing by first intention it is, however, a well-marked structure, and is something more than a mere ring of new tissue enclosing the ligature, for, even in these cases, it extends up and down the vessel for a short distance. In large vessels, it forms a callus of considerable size, even when the healing of the wound has been rapid. This is shown in the vessels of the horse (Frontispiece), and also in those of man, notably in the subclavian artery, and in a specimen of the carotid artery prepared in such a way as to show the callus.

While the callus has been enlarging, its cells have infiltrated the outer fibres of the adventitia, and they gradually disintegrate and absorb the bundle of fibres held by the ligature, which now lies enclosed by a wall of granulation cells on all sides. The walls of the vessel have now completely separated from one another, but form at each end a firmly closed cul-de-sac, which still shuts out all external growth. (Frontispiece, Fig. I.) As two weeks have elapsed by this time, it is not unnatural that observers should have supposed that union had taken place, but in reality we are only approaching the beginning of the second stage of the process. The ligature, which is now entirely independent of the vessel, becomes encysted; or, if it be made of material that is easily absorbed, is found infiltrated with granulation cells, or it may have disappeared entirely. Under the old system of leaving one end of the ligature protruding from the wound, it would by this time have become sufficiently detached from the tissues, which it had originally surrounded, to enable the surgeon to withdraw it, provided the callus did not hold the knot too firmly in its grasp. Occasionally the suppuration would be sufficient to discharge it through an open sinus, as seen in the Frontispiece (Fig. II.). Strictly speaking, the ligature does not ulcerate through the walls of the vessels, as an elastic ligature does through tissues which it encloses. Having constricted the adventitia into a tendinous-like band at the moment of its application, it remains encircling this tissue like a ring upon the finger. The softening of the vessel-wall is accomplished by the granulation cells, and this process is really more marked after the ligature has separated entirely from the vessel. The sole function of the ligature, after it has closed the vessel, is to keep the walls from expanding until they are held sufficiently firmly by the callus.

During the third week, the granulation tissue has so infiltrated the inner and middle coats of the vessel at the point where they were drawn together by the ligature, that they retract slightly, and allow the granulation tissue to insert itself between them, and to invade the interior of the vessel. This is a somewhat critical period in the process of repair, for, if there has been a softening of the callus by suppuration, it may not be sufficiently strong to restrain this retraction and the thrombus may be forced out (Fig. 14) unless it be firmly secured to the vessel-wall by the process alluded to above.

The second stage begins somewhat earlier in dogs than in man or horses, and varies greatly according to the amount of inflammation. When the latter is severe it may already have begun at the end of the first week (Fig. 8). It is not, however, until the end of the first month that we see the walls of the vessel open nearly to their original distance from one another, and the granulation tissue invading in mass the interior (Figs. 11 and 16). As this growth of tissue pushes forward, it infiltrates the lower layers of the thrombus, as in Figure 16, or it pushes the clot before it, as in Figure 11:—The greater portion of the thrombus is, however, eventually infiltrated by these granulations, which, later, are seen communicating directly with the lumen of the vessel (Fig. 18). In cross section the relations of the two are well shown (Fig. 17), the newly formed tissue appearing honey-combed by blood-spaces, giving it a cavernous structure. Viewed in long sections, the internal growth extends up on either wall somewhat farther than in the centre, and generally a longer distance on one side than on the other. (Fig. 16).

The appearance of the vessel at this period is striking (Frontispiece, Fig. II). The walls have opened widely, and are imbedded in a mass of newly formed tissue, which surrounds and is inserted between them. On the upper surface of the inner growth or callus, the remains of the thrombus are usually seen covering the granulations, like a scab.

Where the thrombus has been a large one it appears, to the naked eye, not to have changed, but an examination with a lower power discloses the fact that most of the blood corpuscles have disappeared, and that granulation-cells and fibres have taken their place. The thrombus has, according to the old view, become organized; or, according to more modern ideas, has been absorbed. It has served as a good neutral medium for the granulation cells to grow in, and has acted as a sort of model, giving to the new tissue, to a certain extent, its outward form. This tissue of the internal callus is rich in capillaries which grow in with the tissue, taking their origin from the vessels of the external callus; they ramify in the granulations, but do not as yet communicate with the lumen of the vessel. In some specimens it was possible to follow their development (Plate IV), which had all the appearance of being of that variety known as the intercellular, the spindle-shaped cells grouping themselves in bundles, and forming channels which communicate directly with the capillary system.

The time when the capillaries communicate with the lumen has

10

been a subject of much discussion. It is probable that, in many cases this occurs to a limited extent only, as the majority of them have simply a temporary existence. In larger cicatrices, however, (Fig. 27), the new tissue is rich in fine capillaries and arterioles, which communicate freely with the blood spaces. In such cases communication is not probably established until the tissue has differentiated to such an extent as to become more nearly like the permanent cicatricial tissue. This does not occur until the beginning of the final or third stage; that is, at the end of the second or the beginning of the third month.

Figure III. of the frontispiece, although taken from a preparation of four months, an unusual period, serves well as an illustration of the early part of the final stage of the process of repair. The external callus has already entered upon a retrograde change. While the process of absorption is taking place without the vessel, a process of differentiation of the cell-elements is taking place within. Already we see numbers of spindle-shaped cells with staff-shaped nuclei, particularly near the upper surface of the granulation tissue, or that nearest the lumen.

The ligature at this period may have been cast off, or it may have been absorbed, or it may remain encysted in the callus. Frequently the catgut ligature, and the cotton also, will have disappeared long before this. The silk ligature has been observed as long as four months after its application. No experiments were made to test the different materials commonly used for this purpose.

The third and final stage comprises that period during which the provisional structures are absorbed, and the permanent cicatricial tissue is developed in their place. The cicatrix is situated between the ends of the vessel-walls, which are still separated from one another, and is composed of three constituent parts: the endothelial, the muscular, and the connective tissue portions. In small cicatrices which occupy only a short segment of the interior of the vessel, these tissues are arranged in layers, the endothelium covering the muscular fibre, beneath which lies the connective tissue, which is continuous with the cord. (Frontispiece, Fig. IV. and Fig. 20). In long cicatrices, such as are seen in the largest vessels in man, we have a more complicated arrangement The lumen is obliterated for a considerable distance by a tissue filled with vascular spaces, the muscular cells being arranged chiefly in circular bands around these spaces. Here a condition obtains more nearly resembling that seen in the arteries of amputation stumps, or, as far as it goes, in endarteritis obliterans.

A large portion of this tissue consists of a delicate mucous tissue with branching cells, round cells, and delicate fibres, and filled with capillaries and arterioles, forming a system which communicates, here and there, with the large cavernous blood-spaces. These latter ramify throughout the whole length of the cicatricial tissue, but do not project beyond the slightly incurved walls at the end of the vessel. (Fig. 27). Here the tissue comes in contact with the bundles of fibrillated tissue of which the cord is composed. The muscular fibres are massed chiefly near the open lumen of the vessel and around the blood-spaces, a few of them being arranged longitudinally.

The muscular cells were subjected to a great variety of tests to determine their true character. In dogs a number of specimens were cut transversely, and here they appeared as round cells, (Fig. 24.) In Figure 25 they are arranged in the rent in the mediæ, in circular bands, and again show in cross section. In other specimens they are seen to emerge from the clefts in the lamina elastica, and to be directly continuous with the muscular cells of the media, whether the vessel be studied in longitudinal (Fig. 28) or in cross sections (Fig. 23). In the latter drawing, the muscular cells of the media are seen on the right hand, and the newly formed cells in the cicatrix on the left. In longitudinal sections, the long staff-shaped nuclei are brought out distinctly in sections mounted in Canada balsam (Figs. 22 and 28), and, in Figures 21 and 23, the bodies of the cells show well in glycerine preparations. In Figure 23, the cross section is taken at a point nearly corresponding in Figure 20, to the line V. A great variety of staining fluids were used, all of them bringing out the characteristic nuclei. It has been maintained that cells of this appearance may be seen in cicatricial connective tissue, but there seems nothing improbable, or contrary to nature, in the hypothesis that they are derived from muscular tissue. The proliferation of the muscular cells of the media in the walls of the arteries of horses was observed in the specmens recorded in the series of experiments on horses. In dogs, the growth of these cells into the cicatricial tissue has just been alluded to, and in man the cells of the cicatricial tissue have the same marked type. Muscular cells are formed in large numbers around vessels of considerable size, in the obliterating tissue of arteries, and a similar arrangement of such cells is seen in the new tissue formed within the walls of the hypogastric artery. In the aortic cicatrix, and in the cicatrix of the pulmonary artery, at the ends of the ligamentum arteriosum, we see masses of such cells directly continuous with the muscular coats of those vessels. Finally, the many excellent opportunities to compare these cells with the muscular cells of the vessel-walls in immediate contact with them, leave no reasonable doubt as to their identity.   The new formation of muscular fibre is not an unusual occurrence.   We see cells of this nature formed in abundance wherever new arterioles grow in inflammatory tissue.   The walls of the uterus furnish a physiological example of the development of muscular cells on a large scale, and they are also seen forming a very considerable portion of the morbid growths in the fibro-myomata.*

The shape of the cicatrix in the interior of arteries, as well as its size, varies considerably.   In all cases it is so disposed within the vessel that there shall be a gradual narrowing of the lumen until it becomes the size of an arteriole.   In moderately large vessels this is accomplished without great difficulty, but in full-sized vessels the cicatrix is elongated and composed of cavernous tissue, by means of which the blood-current is allowed to circulate in a system which is the equivalent of an elongated vessel gradually narrowing in calibre, and coiling up in a confined space.   The strain upon the cicatrix is thus probably greatly relieved.   The shape of the cicatrix is also such as to adapt the lumen to the diminished blood-supply which now is sent in this direction.   This supply will depend upon the proximity of a large branch.   If there be no branch in the immediate vicinity, the cicatrix will extend for some distance farther up the sides of the vessel than in the centre (Frontispiece, Fig. IV.), and a small arteriole will take its origin from the apex of the cul-de-sac thus formed.   If a large branch be close at hand, one horn of the cicatrix will extend up on the opposite side, adjusting the size of the current to the needs of the collateral circulation, so that no abrupt change in the diameter of the circulating column of blood can occur (Fig. 26).   The strain brought upon the wall of a vessel at the point of a narrow outlet, from a large arterial blood-space which had no other exit, would be too great.   We find in fact that the exit of all large vessels from the aorta are supported by an extra development of muscular tissue, so placed at the angle in the wall as best to support the pressure brought to bear upon it.   The cicatrix is said to be asymmetrical, or symmetrical according to these varying conditions.   It is probable that no cicatrix is perfectly symmetrical,

---

* For a recent observation of this kind, see Amer. Jour. Med. Sci., April, 1886, p. 511.

for the neighborhood of a branch or branches will affect the size of one of its horns, more or less, in almost every case.

The cord which unites the two ends of the vessel varies greatly, of course, in length. When a vessel has been tied between powerful branches, as certain parts of the subclavian, the ends of the vessel will retract but a short distance from one another, and but little of the vessel-wall will be absorbed. On the other hand, if the ligature be placed upon a long branchless trunk, like the common carotid, the cord will be unusually long. In the latter case there has been, not only a retraction of the ends of the vessel, but also an actual absorption of portions of the vessel-walls, which have now become useless. (Fig. 27). The study of the behavior of vessels in this respect, not only after ligature in man and animals, but also after closure of the hypogastric artery, confirms this view. The cord does not, however, represent the fibrous remains of the vessel-wall, but the final stage of development of the tissue of which the external callus is composed. The greater portion of this structure, is, like the internal callus, absorbed and the cord may be regarded as its cicatricial remains.

The cicatrix of an artery, after ligature in its continuity, bears a close resemblance to the cicatrix in the aorta and in the pulmonary artery, marking the former openings of the ductus arteriosus. We find here the central arteriole surrounded by a delicate muscular tissue, enclosed between the ends of the media, which terminate abruptly a short distance from one another. The comparison is interesting as throwing light upon the nature of the tissues involved in the new cicatrix.

The appearances observed in the interior of an artery of an amputation stump do not differ essentially during the first week from those seen after ligature in continuity. A slight difference is noticed in the incurvation of the media, in vessels which have been seized at the end by forceps, the adventitia being drawn forward over the rolled-up media. This is not always seen, however, in amputation stumps. At the end of this time, there is a marked divergence in the two processes. In the interior of the stump of the amputated artery, there is a thickening of the intima for a considerable distance from the point of ligature. Probably this change can be followed to the neighborhood of the origin of the vessel. By the end of the third week, the lumen of the vessel is greatly diminished in calibre. The character of this growth is shown well in the drawings (Figs. 5 and 15). It consists of a myxomatous tissue containing spindle-shaped and round cells; in some instances small blood-vessels are seen as early as the third week.

The size of the thrombus varies greatly, and not always according to the proximity of a branch, or the amount of sepsis. The tenacity with which the growth from the walls, or through the walls, will hold a thrombus, even when the end of the vessel has been opened by suppuration, is shown in Fig. 14. After withdrawing the needle in acupressure the ends of the vessel may open in this way, but the firmness with which the clot is held prevents its expulsion. There is probably some infiltration of the clot with wandering cells, and also a growth from the intima, which gives this security in these cases.

In the completed process of cicatrization we find a condition apparently very different from that above described in ligature in continuity, but which is, in reality, a further elaboration of the same process, by which the vessel is adapted to the needs of the circulation. The entire lumen of the vessel must be narrowed for this purpose, and, since this cannot be effected by a simple contraction of the walls, the cicatrix is continued up through the lumen to the point of origin of some large collateral branch. An obliterating endarteritis has taken place, and in the new tissue we see the lumina of several smaller vessels. This condition resembles that seen in the carotid after ligature in continuity (Fig. 27), but it extends through a greater part of the vessel. A study of the tissues in the two cases shows no essential differences; we have, in both, the internal blood-spaces or vessels, the surrounding bands of muscular fibre, and the connective tissue formation. We see in the latter case also, although in a different shape, the gradual tapering of the lumen of the vessel until it terminates in a fine arteriole which communicates with capillaries. In the former case the lumen is not coiled up in a small space, but is extended out through the whole length of the stump. The true character of the process of repair could not be well shown by a dissection-preparation of the vessel of the stump, for in the femoral artery, mentioned above, the outward form still remains, although the lumen has been altered into a number of intertwining vessels. The object would be best accomplished by a corrosion-specimen, which would show only the respective calibres of the vessels, in which case we should probably see the main artery, soon after it entered the limb, breaking up into a spray of smaller vessels, which, with collaterals of a similar size, would be distributed equally to different parts of the stump.

The repair after closure of the hypogastric artery is precisely similar, like conditions as to circulation prevailing here. We see the early thickening of the intima, and the final obliterating tissue pierced by a blood-vessel, with new muscular walls. In the same category can be placed the vessels which have become obliterated spontaneously by a growth within their walls. The possible origin of this growth, in response to a demand that has been made upon the walls of the vessels, to meet the necessities of a diminished blood supply, has already been suggested. Such a change extending so far from the original seat of the traumatic inflammation, should be regarded as formative rather than inflammatory. This example serves to emphasize the difference between repair and inflammation.

The old view that this obliteration is the result of a " thickening of the intima," that is, a growth solely from the cells of the intima, is untenable in the light of these investigations. The bulk and strength of the arterial wall lies largely in its middle coat, and, on *a priori* grounds alone, it seems improbable that this coat should take no part in processes which are of so vital import to its future usefulness. In the experiments given here we see coming from this layer actual growths, and a proliferation of the muscular cells which it contains, and it is highly probable also that a considerable amount of connective tissue may be derived from this coat. It is somewhat remarkable that in all the investigations we have quoted in the historical summary no mention is made of such a possibility. Herein lies the secret of the security of the ligature, and of the fact that after this operation no aneurismal dilatation of the vessel takes place. In wounds of the walls of arteries the circulation is re-established long before a durable cicatrix, which requires months for its formation, has been developed, and the tender and non-contractile tissue is easily dilated by the pressure of the column of blood. The powerful provisional structures, which develop after ligature, are sufficient to control this pressure until the young cicatrix has grown strong enough to do its work without their aid.

Since the ligature became an important feature of operative surgery, a great deal of ingenuity has been expended in the effort to occlude the vessel without subjecting it to the dangers of secondary hemorrhage. For this purpose, it was at one time thought advisable to apply several ligatures to the vessel, side by side, in order to bring a great portion of the walls in contact with one another; later, broad and flat ligatures were also employed for the same purpose. Attempts were also made to imitate more closely the obstruction of the circulation by compression by means of the application of two ligatures, near together, which held the walls gently in contact without rupturing the inner coats. Another supposed improvement over the simple ligature in continuity consisted in the application of two ligatures, and the division of the vessel between them, in order that the tendency of each end to retract might not tear the two fragments apart before they had been effectively sealed by the healing process. In acupressure and torsion a new principle was touched. The design was to avoid, by these methods, the presence of a substance which interfered with repair. The ligature as formerly applied produced conditions similar to those which prevail in bones subjected to compound fracture. The long end of the thread prevented closure of the wound, and favoured suppuration at a point where union by first intention was the great object to be attained. If both ends were cut short, it was supposed that the knot would eventually work its way out, like any other foreign body, and, under the conditions which prevailed before aseptic precautions were employed, this was usually the case. Even with the advantages which antiseptic surgery gave, it was thought necessary to employ some material which could be subsequently disintegrated and absorbed. The animal ligature was then introduced with this object in view. We now know that both silk and hempen ligatures can become either encysted or absorbed; in other words, they can be so applied as not to interfere with the healing process.

Provided the ligature be adjusted so as to obstruct permanently the flow of blood through the vessel, it is manifest, from the observations which have been described, that a destruction of a certain portion of the vessel-walls, and a retraction of the ends of the vessel, must eventually take place, no matter what the nature of the material may be, or how it be applied.

The prime object, therefore, to be obtained, is to employ such methods as will interfere as little as possible with the natural sequence of events which follow one another during the process of repair under the most favorable conditions. When the ends of the vessel are once sealed by the formation of an external ring or callus, and the rest of the wound is promptly healed by first intention, so that this growth shall not be prematurely broken down by suppuration, all danger of hemorrhage is avoided. The rules of antiseptic surgery supply us, therefore, with a more certain method of securing this desirable result than any other plan which, up to the present time, has been proposed.

APPENDIX.

METHODS.

ALL fresh specimens were placed immediately in Müller's fluid, and kept there during a period varying from two days to three weeks, according to the size of the vessel. After being thoroughly washed in water, they were placed in strong alcohol.

In case the thrombus was the object of study, the vessel was opened before placing it in any preservative fluid. The material found best for imbedding was celloidin, paraffine frequently rendering the thrombus so tough as to turn the edge of a razor. In imbedding in celloidin care should be taken to allow the material to flow into the lumen of the artery: the vessel should, therefore, be cut open as close to the point of ligature as possible, to permit the bubbles of air to escape; otherwise, the edge of the wall opposite an air bubble will bend under the razor, and be cut obliquely in consequence.

Longitudinal sections proved to be much more instructive, as a rule, than transverse sections. No attempt was made to arrange the longitudinal sections in serial order, for a little experience enables one to distinguish between those which come from the centre of the specimen, and those from the upper and lower surfaces. An inexperienced observer can, however, be easily deceived by an apparent thickening of a particular portion of the wall, due to the obliquity of the section. This danger is avoided by selecting the central sections.

In making cross sections, more attention should be paid to preserving a knowledge of the order in which they have been made. In specimens taken from ligatured arteries, it was usually sufficient to separate the proximal from the distal portions before imbedding for cross sections; but where an entire vessel was examined, like the hypogastric artery, it was found necessary to divide the specimen into seven or eight different segments, which were numbered to correspond to the figures on a diagram preserved for the purpose. Some of these segments were cut into longitudinal, and some into transverse sections. In studying the ductus arteriosus, the best results were

obtained by cutting the specimens into halves, whether for the purpose of making longitudinal or transverse sections.   Hæmatoxyline
was the staining fluid chiefly used, although many of the aniline dyes
were experimented with in studying muscular fibre.   Very ornamental specimens were obtained by the double staining with eosine and
hæmatoxyline.   The best material for mounting proved to be Canada
balsam, owing to the transparency of the thrombus in this medium.
Glycerine was also used, but chiefly for detail studies of cell-
structures.

The student's Hartnack-microscope was the instrument chiefly
used, and the numbers given in the explanation of the figures are
those of the lenses used with this instrument.   The highest power
ordinarily employed was the nine-immersion lens, although Zeiss'
oil-immersions were occasionally used.   High powers were, however,
rarely needed.   The letter t, used in describing the power, refers
to the slight increase of power produced by the drawing out of the
tube.   In carrying out aseptic precautions in the ligature of arteries
in dogs, the spray was usually employed although equally good results were obtained without it.   The animal should be washed the
day before the operation, and before making the incision the integu-
ments over the vessel should be shaved and thoroughly scrubbed
with one-to-forty carbolic wash.   It is well also to dust on iodoform
powder at this time.   All usual precautions in the care of instruments
and hands should, of course, be taken.   A few strands of carbolized
gut should be made to serve as a drain to the more superficial portions of the wound, and catgut sutures should be employed.   The
material used for the ligature varies, of course, according to the de-
signs of the operator.   The wound and the neighboring parts should
be thoroughly dusted with iodoform, which is peculiarly well adapted
for these cases, as the hair retains the powder quite as well as it
does dirt.   A borated cotton-dressing can be easily retained, either
around the neck or the groin by a free use of the spica bandage,
and, when properly applied, it will be found at the end of the week,
the length of time it was usually left on, not to have been disarranged
in the least.   As an additional precaution, it is well to dust in a little
iodoform daily, about the edge of the dressing, and the animal
should be kept shut up alone, during the healing process.

The very marked difference in the healing of wounds treated in
this way and of those in which no antiseptic precautions were taken,
together with the ease with which antisepsis was carried out, are strong
proofs in favor of the usefulness of this method of treatment of the

wounds of animals. The use of iodoform is strongly recommended, owing to the peculiar tenacity with which it clings to the hair.

The wounds made in experiments on horses were not treated in this way, and healing was protracted, and occasionally inflammation was excessive. The use of iodoform before, during, and after an operation on this useful animal, ought to be of great value in obtaining satisfactory results.

DESCRIPTION OF PLATES.

FRONTISPIECE.

I.—Carotid artery of horse two weeks after ligature. The specimen shows the external callus and the thrombus: in this specimen the proximal thrombus is white, for what reason does not appear. The walls of the vessel are continuous around the base of each thrombus, and they appear to have united by first intention, which is not the case.

II.—Carotid artery of horse two months after ligature, showing the external and internal callus, and the open ends of the vessel. The second stage of repair.

III.—Carotid artery of horse four months after ligature. The third stage of repair. Absorption of the external callus.

IV.—The external iliac artery. Man. One hundred and thirty days after ligature. Termination of the third stage. Formation of muscular cicatrix.

PLATES.

A. Adventitia. B. Blood-clot. C. Ligament. D. Pigment.
E. Lamina Elastica. F. Muscular cells. G. Granulation tissue.
H. Granulations. I. Intima. K. Endothelium. L. Ligature.
M. Media. N. New Growth. P. Periadventitia.
S. Blood Spaces. T. Thrombus. V. Vessel.
The number of each Hartnack objective and eye-piece is given. The letter t indicates that the tube of the microscope was drawn out.

PLATE I.

Fig. 1. (2x2xt).—Femoral artery of dog, two days after ligature. The action of the ligature on the walls is shown: wandering cells are entering the thrombus, and granulation tissue surrounds the knot.

Fig. 2. (2x2).—Carotid artery of dog four days after ligature. The upper end, the proximal portion, is distended: the distal end is contracted. The white corpuscles have not wandered in, but belong to the clot.

PLATE II.

Fig. 3. (9x2xt).—Proliferation of the endothelium in the distal portion of Fig. 2. Mother-cells, spindle-cells, endothelium and blood-corpuscles are shown.

Fig. 4. (2x2).—Brachial artery of man, two hours after laceration. The external and internal thrombus are shown, and the curling of one side of the vessel.

Fig. 5. (9x3xt).—Tibial artery, from amputation stump of man, three weeks after operation. Growth of cells in the intima. The double elastic lamina of the tibial artery is also shown.

Fig. 6. (7x3xt).—Carotid artery of dog, one week after ligature. Rupture of elastic lamina at a point some distance from the ligature. Growth of cells from the media and intima.

PLATE III.

Fig. 7. (9x3xt).—Femoral artery of dog, ten days after ligature. Growth of endothelium at the ruptured edge of the inner coats at the point of ligature. Blood plaques at B.

Fig. 8. (2x2).—Femoral artery of dog, nine days after ligature. Opening of the distal end of the vessel. Ingrowth of cavernous granulations. No clot shown.

Fig. 9. (2x2).—Same as Fig. 7. The walls have not separated, but granulation tissue is infiltrating and softening them. The thrombus is attached by a pedicle.

Fig. 10. (9x3xt).—Same as Fig. 7. A growth of the endothelium over a portion of the wound in the inner wall. Blood plaques at B.

PLATE IV.

Fig. 11. (2x2).—Femoral artery of dog fourteen days after double ligature. Opening of the ends of the various portions of the vessel, with ingrowth of granulations.

Fig. 12. (9x3xt).—Formation of capillary vessel beween rows of spindle-shaped cells at point V in Fig. 11.

Fig. 13. (9x4xt).—The same as Fig. 12.

PLATE V.

Fig. 14. (2x2).—Tibial artery from amputation stump of man, three weeks after operation. Thrombus protrudes from the open end of vessel, owing to suppuration in wound. Growth in intima. Infiltration of the walls.

FIG. 15. (7x3xt).—Growth in intima of Fig. 14, under a high power. New formed blood-vessels or blood-spaces.

PLATE VI.

FIG. 16. (2x2).—Femoral artery of dog, one month after ligature. Proximal portion. The open end of the vessel permits the growth of granulation tissue into the interior. Near the lumen granulations are seen. The clot is partially absorbed.

FIG. 17. (2x3xt).—Femoral artery of dog, one month after ligature. Cross section of distal portion showing the growth of granulations into the lumen, and the formation of blood-spaces between them.

PLATE VII.

FIG. 18. (7x3xt).—Detail study of granulations in Fig. 16, showing the hyaline structure and their endothelial covering, also the blood-spaces between them.

FIG. 19. (2x3xt).—Study of a portion of the proximal end of Fig. 17, showing a rupture of the elastic lamina, and a growth into the vessel from the media. This growth formed a diaphragm a short distance above the apex of the thrombus, traces of which still remained.

PLATE VIII.

FIG. 20. (4x3xt).—Femoral artery of dog, three months after ligature. The appearance of the final cicatrix. The muscular portion (F) of the cicatrix contrasting with the connective tissue portion below it. The asymmetry of the cicatrix is also shown.

FIG. 21. (9x4xt).—A study of the muscular portion of Fig. 20, the specimen being mounted in glycerine. The cells on the left form the endothelial covering.

FIG. 22. (9x4xt).—A similar study of the deeper layers, near the elastic lamina. Specimen mounted in Canada balsam.

PLATE IX.

FIG. 23. (9x4xt).—Cross section taken on about a level with the line V in Fig. 20. On the right is the media; on the left, the new muscular cells growing from it into the cicatrix. Glycerine preparation.

FIG. 24. (7x3xt).—Cross section from the cicatrix in the carotid artery of a dog, four months after ligature. The muscular cells here look like round cells. The section is taken about opposite V in Fig. 26. Canada balsam preparation.

FIG. 25. (4x3xt).—Wound made in the walls of carotid artery of a dog, by applying ligature, and removing it immediately. Speci · men removed three months after ligature. The wound is partly filled by a growth of new muscular tissue.

PLATE X.

FIG. 26. (2x2).—Carotid artery of dog, four months after ligature. The drawing shows the shape of the cicatrix as modified by the presence of a branch.

FIG. 27. Natural size. Diagram of the left common carotid artery of man, four years after ligature. The shape and cavernous structure of the cicatrix are shown.

FIG. 28. (9x3xt).—Longitudinal section of a portion of the cicatricial tissue in Fig. 26, of which Fig. 24 is a cross section. The endothelium is seen on the surface and below the muscular cells, which are growing through a rent in the lamina from the media. Canada balsam preparation.

PLATE XI.

FIG. 29. (2x2xt).—The aortic end of the ligamentum arteriosum, at forty-two years, in the human subject. The ends of the media aortæ are seen slightly separated. A central arteriole follows the axis of the ligament, and is surrounded by new muscular cells. On a level with C, outside, circular muscular fibres are seen, and, in the two inner layers, longitudinal muscular fibres. Around the point of origin of the arteriole is a growth of elastic tissue. Section taken in a vertical plane.

FIG. 30. (2x2xt).—Aortic end of the ligamentum arteriosum, at forty-two days. In the space opposite T a thrombus existed. Around this space are seen the walls of the ductus arteriosus, in a state of hyaline degeneration. The aortic cicatrix is forming on the left. The section is taken in a horizontal plane.

FIG. 31. (2x2xt).—Aortic end of the ligamentum arteriosum at thirty-eight years. The usual depression in the aorta is wanting; and this peculiarity is due to the presence of a large amount of elastic tissue, which bridges over the space between the edges of the media. The central vessel appears to come from the pulmonary artery. Patches of calcification are seen.

PLATE XII.

FIG. 32. (7x3xt).—Cross section of the hypogastric artery of an adult; showing the new formed muscular and elastic layers, within the lumen of the old vessel.

FIG. 33. (2x2.)—Femoral artery, taken from a stump at least fifteen years after amputation. A dissecting room subject. The cadaveric changes prevented staining; but the specimen shows well the new tissue formed within the old walls, the inner borders of which are marked by the elastic lamina. Spots of calcification are seen.

FIG. 34. (7x3xt.)—Hypogastric artery of a monkey. The new vessel formed within the old shows with great distinctness. There is no new elastic lamina.

FIG. 35. (4x3xt).—Endarteritis obliterans. Cross section of the tibial artery, in human subject. The media is much altered and calcified. The lamina elastica is broken. The new tissue contains a large vessel with new formed muscular wall.

11

BIBLIOGRAPHY.

1. B.C. 1500. Súsrutas Ayur-Védas : Id est medicinæ systema, a venerabili D'harantare demonstratum a Súsruta discipulo compositum. Hessler, Erlangen, 1844–50.

2. B.C. 400. Hippocratis medicorum omnium facile principis opera omnia quæ exstant autore Anutio Faesio. Frankfurt, 1621, ii., 1194. The genuine works of Hippocrates, translated from the Greek, with a preliminary discourse and annotations by Francis Adams, LL.D., London. Sydenham Soc., 1849.

3. B.C. 30. Aurelius Cornelius Celsus, on medicine, in eight books, Latin and English. Translated from L. Tarja's edition, by Alexander Lee, London, 1831, also by James Grieve and George Futvoye, London, 1837. A translation of the eight books of Aul. Corn. Celsus on medicine, 3d ed., by G. F. Collier, M.D., London, 1843. Book V., cap. xxvi., 187. Book VII., cap. xix., 209.

4. A.D. 360. Oribasius Συναγωγαὶ ἰατρικαί, œuvres d'Oribase, par Bussemaker et Daremberg, Paris, 1851–62. Græcorum chirurgici libri, etc., editi ab Antonio Cocchio, Florentiæ, 1774. Oribasii cap. xiii., de amputandis partibus ex Archigene, p. 155. Cap. xiv., Heliodori de extremis membris abscindendis, p. 156.

5. 1st and 2d Cent. Analecta historico-medica de Archigene medico et de Apolloniis medicis, etc., Lips., 1816. 4. C. F. Harless.

6. 1st and 2nd Cent. De Heliodori veteris chirurgi fragmentis. Diss. Gryphiso, 1846. 8. T. Lenz.

7. 3d Cent. Claudii Galeni opera omnia. Ed. Kuhn, Lipsiæ, 1827–30. De Methodi Medendi V., cap. iii., p. 378.

8. 6th Cent. Aetius βιβλία. ἰατρικὰ ἑκκαίδεκα. The works of Aetius translated into Latin by Cornarus and Montanus, Basil, 1533–35, fol. p. 47 (a quotation from Rufus).

9. 7th Cent. The seven books of Paulus Ægineta. Translated from the Greek, etc., by Francis Adams, London, 1846. Sydenham Soc., pp. 130, 128.

10. 10th Cent. Abubetri Rhazæ Maomethi liber continens Basileæ, 1544. Lib. I. VII., de fluxione sanguinis ex vulneribus.

11. 10th and 11th Cent. Avicenna. Arabum medicorum principis ex Gerardi Cremonis versione, et Andreæ Alpagi Bellumensis castigatione. Per Fabium Paulinum Venetiis apud Juntas, 1595.

12. 12th Cent. Averrhoes. Liber de medicina 64 ff., fol. Venetiis, 1490, also Avenzohar, of same place and date.

13. 12th Cent. Avenzohar (abhumeron) incipit liber theicrisi dahalmodana vahaltudabir, etc. per Joannem de Forlivio et Gregorius fratres. Anno salutis, mcccclxxxx.

14. 12th Cent. Albucasis de chirurgia, arabice et latine. Cura Johannis Channing, Oxford Clarendon Press, 1778. La chirurgie d'Abulcasis, Traduction française du Dr. L. Leclerc.

15. 13th Cent. Averrhoes. Incipit liber de medicina Aueroys qui dicitur Coliget, etc., anno 1482, impr. Veneciis per Laurent de Valentia et socios.

16. 1295–6. Lanfranchi's surgery, also Lanfranci. Practica quæ dicitur ars completa totius chirurgiæ. Venet., 1490 f., also Coll. Chir. Ven., 1519.

17. 1363. Cyrurgia Guidonis de Cauliaco et Cyrurgia Bruni, Feodorici, Rolandi, Lanfranchi, Rogerii, Bertapalie Venetijs per B. Venetum de Vitalibus, 1519.

18. 1450. Leonardo Bertaplagia. Cyrurgia S. Recollectæ super quarto Avicennæ. Venet, 1498, de vulneribus, c. 20.

19. 1497. Dis ist das buch der Cirurgia, Hautwürckung der Wundartzney von Hieronymus Braunschweig. Augsburg, also English translation London, 1525.

20. 1506. Angelus Bolognini. De cura ulcerum exteriorum. Venet.

21. 1514. Marianus Sanctus de Barletta (s. Barolitamus) compendium in Chirurgia (Rom?).

22. 1543. Johannis Tagaultius. De chirurgica institutione libri quinque. Paris.

23. 1552. Alph. Terrius. De Sclopetorum sive Archibusorum vulneribus libri tres, etc.

24. 1575. Les œuvres de M. Ambroise Paré avec les figures et portraicts tant de l'anatomie que des instruments de chirurgie et de plusieurs monstres. Paris, 1575, 1579, 1585, 1598, etc. ii., 224.

Ambroise Paré. Œuvres complètes, revues et collationnées sur toutes les éditions, par J. F. Malgaigne. Paris, 1840. The works of Ambrose Parey, London, 1691, 306.

25. 1612. Jacques Guillemeau. Les œuvres de chirurgie. Paris.

26. 1652. Wilhelm Fabriz von Hilden. Wundarzneikunst. Frankfort a. M. (Wilhelm Fabry of Hilden or Fabricius Hildanus).

27. 1660. Leonardi Botalli. Opera omnia, Lugduni Batavorum, p. 790.

28. 1672. Richard Wiseman. A Treatise of wounds, also experiments made at London for staunching the blood of arteries and veins. Phil. Trans. Apr. ii., 17, 1673.

29. 1674. Morel. Mém. de l'acad. roy. de Chir. ii., 390.

30. 1684. Cornelis van Solingen. Manuale operative der Chirurgye. T' Amsterdam.

31. 1702. Barthélémy Saviard. Nouveau recueil d'observations chirurgicales.

32. 1707. Pierre Dionis. Cours d'opérations de chirurgie demontrées au Jardin du Roi. Paris.

33. 1718. Lorenz Heister. Chirurgie. Nüremberg.

34. 1720. R. J. Garengeot. Traité des opérations de chirurgie. Paris.

35. 1731. Jean Louis Petit. Dissertation sur la manière d'arrêter le sang dans les hémorrhagies. Mémo. de l'acad. royale des sci. Paris, also ibid. 1732 and 1735.

36. 1734. Alexander Monro. Remarks on the coats of the Arteries, their diseases, and particularly on the formation of an aneurism. Med. Essays and Obs. Edinb., ii. 264.

37. 1736. François Sauveur Morand. Sur les changements qui arrivent aux artères coupées, ou l'on fait voir qu'ils contribuent essentiellement à la cessation des hémorrhagies. Mém. de l'Acad. royale des sci., Paris.

38. 1739. Samuel Sharp. A treatise on the operations of surgery, etc. London.

39. 1742. Henri François Ledran. Traité des opérations de chirurgie. Paris.

40. 1753. Antoine Louis. Seconde mémoire sur l'amputation des membres. Mém. de l' Acad. royale de chir. ii., 397.

41. 1760. Claude Pouteau. Mélanges de chirurg. Lyon.

42. 1763. Thomas Kirkland. Essay on the method of suppressing hemorrhages from divided arteries. London.

43. 1770. Charles White. Cases in surgery, with remarks, to which is added a treatise on the ligature of arteries, by J. Aitkin. London.

44. 1773. William Bromfield. Chirurgical observations and cases. London.

45. 1782. August Gottlieb Richter. Anfangsgründe der Wundarzneikunst. Göttingen.

46. 1792. Benj. Gooch. Chirurgical works. London.

47. 1793. J. F. Louis Deschamps. Observations sur la ligature des principales artères. Paris.

Observations et reflexions sur la ligature des principales artères blessés. Paris, 1797, 33, 53.

48. 1793. J. Abernethy. Surgical and physiological essays. London.

49. 1794. John Hunter. A treatise on the blood, inflammation and gunshot wounds. London.

50. 1798. Pierre Joseph Desault. Œuvres chirurgicales. Paris.

51. 1799. Antylli, veteris chirurgi, τα λείχανα Nicolaïdes (Panaïota) Diss. Inaug. Halis Magdeb.

52. 1802. Jean Pierre Maunoir. Mémoires physiologiques et pratiques sur l'anévrysme et la ligature des artères. Paris.

53. 1805. J. F. D. Jones. A treatise on the process employed by nature in suppressing the hemorrhage from divided and punctured arteries. London

54. 1806. James Veitch. Edinb. Med. and Surg. Journal, ii., 176.

55. 1812. Mémoires de médecine et de Chirurgie militaire et de Campagnes de J. D. Larrey. Paris.

56. 1815. Joseph Hodgson. A treatise on the diseases of arteries and veins, containing the pathology and treatment of aneurisms and wounded arteries, xix, London. T. Underwood.

57. 1815. John Bell. Principles of Surgery, London, i., 141. Discourse on the Nature and Cure of Wounds. 1800, p. 109.

58. 1817. Anton Scarpa. Memoria sulla ligatura delle principale arterie. Pavia.

59. 1818. Johann Friedrich Meckel. Handbuch der pathologischen Anatomie, ii., Leipsig.

60. 1819. Crampton. Med. Chir. Trans. vii.

61. 1819. Travers. Med. Chir. Trans., vi., 632.

62. 1821. Balthasar Anthelme Richerand. Nosographie et thérapeutique chirurgicales, Paris.

63. 1823. Delpech. Chirurgie clinique de Montpellier, i., 109. Paris et Montpellier.

64. 1824. Boullaud. Archives gén. de méd.

65. 1824. Cooper (Sir Astley Paston) Lectures on the principles and practice of surgery, with additional notes and cases by Frederick Tyrrell. London, 1824-27.

66. 1825. Ribes. Revue médicale française et etrangère, iii.

67. 1825. C. J. M. Langenbeck. Nosologie und Therapie der chirurgischen Krankheiten. Göttingen, iii., 134.

68. 1826. Ebel. De natura medicatrice sicubi arteriæ vulneratæ et ligatæ fuerint Giessen.

69. 1826. Vatel. Sur le thrombus et la phlébite partielle. Jour. prat. de méd. veterin.

70. 1826. Koch Journal von Gräfe und Walther, ix., 560, Berlin.

71. 1826. Gendrin. Histoire anatomique des inflammations, ii., Paris.

72. 1827. Holtze. Diss. de arteriarum ligatura, Berlin.

73. 1827. Rigot et Trousseau. Archives générales, Paris, xiv.

74. 1828. Schonberg. Journal des progres. Paris, xii., 70.

75. 1829. A, Thierry. De la torsion des artères, Paris.

76. 1829. J. F. Lobstein. Traité d'anatomie pathologique générale et spéciale, Paris.

77. 1829. H. S. Levert. Experiments on the use of metallic ligatures as applied to arteries. Am. Jour. M. Sci. iv., 17, 23.

78. 1829. Henry Levert. Journ. von Gräfe und Walther, xiii., 561.

79. 1830. Béclard. Recherches et expériences sur les blessures des artères. Mém. de la soc. méd., viii.

80. 1830. G. J. Guthrie. Diseases and Injuries of the arteries. Wounds and injuries of arteries, London, 1846.

81. 1830. Blandin. Journal hebd. de méd., Paris, Mai.

82. 1830. Andral. Précis d'anatomie pathologique. Bruxelles, ii., 59, 71, 76, 77.

83. 1830. Velpeau. Jour. universelle et Hebd. i., no. 5. Nouveaux éléments de médecine opératoire, Paris, 1832.

84. 1830. C. A. C. Schrader. De torsione arteriarum. Berolini, 1831.

85. 1831. Dietrich. Das Aufsuchen der Schlagadern, behufs unterbindung, etc. Nürnburg. 333.

86. 1831. Pégot. Observations et réflexions sur la torsion des artères. J. univ. et hebd. de med. et de chir. prat., Paris, v. 229.

87. 1831. Bujalski. Journal v. Gräfe und Walther, xv., 402 ibid. Pégot iv., 461, ibid Uso Walter xvi., 355.

88. 1831. Buet. Nouveaux faits sur la torsion des artères et sur le refoulement de leurs membranes. J. compl. de dict. d. sc. med. Paris. xli. 282.

89. 1831. Bédor. Observations et réflexions sur la torsion des artères. Gaz. d. hop., Paris, vi., 46.

90. 1832. Dupuytren. Leçons orales de clinique chirurgicale, Paris.

91. 1832. Manec. Traité de la ligature des artères. Paris.

92. 1832. Uso Walther. Untersuchungen über die temporären Unterbindungen der Arterien, etc. Journal von Graefe und Walther, xvi.

93. 1833. B. B. Bamberger. De variis torsionis arteriarum methodis, Berolini.

94. 1833. T. Brockmuller. De arteriarum torsione, Bonnæ.

95. 1834. Rust. Theoretisch-practisches Handbuch der Chirurgie. Berlin und Wien. xv., 206, 209.

96. 1834. W. Youatt. On the torsion of arteries to arrest hemorrhage in veterinary operations. Lancet, London, ii., 386.

97. 1834. Cruveilhier. Anatomie pathologique. Maladies des veines et des artères, Paris.

98. 1834. B. Stilling. Die Bildung und Metamorphose des Blutepfropfes oder Thrombus in verletzten Gefässen, Eisenach.

99. 1835. D. Van Dockum. Quænam mutationes pathologicæ inducuntur arteriis per ligaturam. Ann. Acad. Rheno-Traject., 1-73.

100. 1839. Stannius. Ueber krankhafte Verschliessung grösserer Venenstämme.

101. 1840. Remak. V. Ammon's Monatsschrift für Medicin, Augenheilkunde und Chirurgie, iii.

102. 1840. Nicolaus Pirogoff. Ueber die Durchschneidung der Achillessehne, Dorpat.

103. 1841. Hasse. Specielle pathologische Anatomie, Leipzig.

104. 1843. Tiedemann. Von der Verengerung und Schliessung der Pulsadern in Krankheiten, Heidelberg.

105. 1843. Vogel. Pathologische Anatomie, 102. Leipzig.

106. 1844. Carl Rokitansky, Handbuch der pathologischen Anatomie. Wien. also 1856, ii., 350.

107. 1844. Amussat, Gazette médicale de Paris. No. 44, 1845, No. 25. Archives Gen. de Méd. Aout. 1829, xx. Recherches experimentales sur les blessures des artères et des veines. 1843.

108. 1845. Virchow. Ueber den Faserstoff. Ges. Abhdlg. 1845, 1846, 1855. Med. Zeit. des Vereins für Heilk. in Preussen, 1847, Sept. Literar. Beilage No. 35. Archiv. f. Path. Anat. i., 272, 1847, Würzburger Verhandlungen 1851, ii., 315. Gesam. Abhdlg. zur wis. Med. Frankfort a M., 1862.

109. 1845. Luigi Porta, Delle alterazione pathologiche delle arterie per la legatura e la torsione, Milano. Delle ferite delle arterie, 1852. Ann. de thérap. No. 7, 1847.

110. 1845. H. Zwicky. Die Metamorphose des Thrombus, Zürich.

111. 1846. J. Lisfranc. Précis de médecine opératoire, ii., 791, Paris.

112. 1851. Reinhardt. Ueber die Metamorphose faserstofliger Exsudate. Deutsche Klinik, No. 36, also, Patholog. anatom. Untersuchungen, herausg. von Leubuscher, Berlin, 1852, 42.

113. 1851. Notta. Observation pour servir à l' histoire de la pathologie du caillot, qui se forme dans les artères à la suite de leur ligature. Gaz. d. Hop., Par. 3, 5, iii., 174, also Recherches sur la cicatrisation des artères à la suite de leur ligature. Gaz. med. de Par., 1850, 3, s. v., 870 ; also Ann. de Med. belge., Brux., 1850, iv., 413, 1851, i., 122, ii., 119 ; also Mémoire sur l'obliteration des artères ombilicales, et sur l'artérite ombilicale Mém. de l' Acad. Imp. de Méd. xix., 1855, Paris. also, Mém. Soc. de Chir. de Par. 1857, iv., 477.

114. 1851. J. Piernas. Experiments with the ligature on animals. N. Orl. M. & S. Jour. viii., 481.

115. 1851. Gerstäcker. De regeneratione tendinum post tenotomiam, Berolini.

116. 1851. Feigel (J. T. A.) Chirurgische Bilder zur Instrumenten und Operations Lehre, vollendet von Textor, Würtzburg.

117. 1852. Henry Lee. Medico-Chirurgical Transactions. Lee and Beale, ibid., 1867.

118. 1852. Thierfelder. De regeneratione tendinum.

119. 1853. Risse. Observationes quædam de arteriarum statu normali atque pathologico, Regiomont.

120. 1854. Butcher, On wounds of arteries and their treatment. Dublin Quarterly Journal, Aug.

121. 1854. Boner, Die Regeneration der Sehnen. Archiv. f. patholog. Anat.,ix.

122. 1857. E. Brücke, Ueber die Ursache der Gerinnung des Blutes. Archiv. f. patholog. Anat., xii.

123. 1858. C. A. Gayet. Nouvelles recherches experimentales sur la cicatrisation des artères après leur ligature, Paris.

124. 1858. Robin. Bulletins de l'Academie de Méd. also 59. Mémoire sur le retraction, la cicatrisation et l'inflammation des vaisseaux ombilicaux. Mém. de l' Acad. Imp. de Méd. xxiv., Paris.

125. 1859. C. Rauchfus. Ueber Thrombose des ductus arteriosus Botalli. Archiv. f. path. anat., Berlin.

126. 1859. L. Buhl. Einige Fälle von Thrombose und Bemerkungen über Atheromatose. Wiener Med. Wochenschrift, No. 34.

127. 1860. J. Dix. On the advantages of acupressure over the ligature. Med. Times and Gazette, London, i., 546. On the wire compress as a substitute for the ligature. Edinb. Med. Jour., x., 207-219.

128. 1860. Cohn. Klinik der embolischen Gefässkrankheiten, Berlin.

129. 1860. J. Y. Simpson. Acupressure, a new method of arresting surgical hemorrhage. Edinb. M. Jour. v., 645. Med. Times and Gazette, London, 1860, i,, 137, 1864 n. s. i., 1, 25 ; 33, 81, 141. Did John de Vigo describe acupressure in the sixteenth century? Brit. M. J. 1867, ii., 145 ; also paper dated 1860, also 1867.

130. 1860. D. Wachtel. Neues Verfahren Zur Hemmung der Blutung, etc. Ztschr. f. Nat. u. Heilk. in Ungarn. Oldenburg, xi. 121.

131. 1861. Ollier. Gaz. Hebd. de Méd. et Chir., 135.

132. 1861. L. H. Jackowitz. Ueber die Acupressur der Arterien, Dorpat.

133. 1861. Hönisch. Historische Nachweis für das hohe Alter der Arterien-Torsion. Allg. Wien. Med. Ztg., vi.,125.

134. 1862. Nélaton. Gaz. des Hôpit., 146.

135. 1862. E. Lanceraux. Gaz. Méd. de Paris, No. 44.

136. 1863. v. Recklinghausen. Ueber Eiter und Bindegewebskörperchen. Arch. für patholog. Anat. etc., Berlin, xxviii., 157.

137. 1863. Theodor Billroth. Die allgemeine chirurgische Pathologie und Therapie, Berlin, 1863 etc., Billroth's Surgical Pathology, New York, 1871, Berlin, Klin Woch. 1871, Chirurgische Briefe aus den Kriegslazarethen, Berlin, 1872.

138. 1864. Holt. Lancet, July 23.

139. 1864. C. O. Weber. Ueber die Vascularisation des Thrombus. Berlin Klin. Woch., 1864. Handbuch der Allg. und spec. Chir. von Pitha und Billroth Bd. 1, 141, Erlangen, 1865.

140. 1865. W. Pirrie. On acupressure. Med. Times and Gaz., London, ii., 5.

141. 1865. R. L. Tait. On the results of temporary metallic compression of arteries. Med. Times and Gazette, London ii., 57, 85, 1866, 1, 197; 335, 1867, i, 332. Letter on ligature, acupressure and torsion. Lancet, London, i., 695, 1869.

142. 1865. August Foerster. Handbuch der speciellen pathologischen Anatomie, Leipzig, ii., 737.

143. 1866. His. Die Häute und Höhlen des Körpers. Basel.

144. 1866. E. Rindfleisch. Lehrb. der patholog. Gewebelehre, Leipzig, 1873, 1875.

145. 1867. E. Célestin. Recherches sur les alterations des artères à la suite de la ligature, Paris.

146. 1867. E. F. Cocteau. Recherches sur les alterations des artères à la suite de la ligature, Paris.

147. 1867. B. Howard. Interesting experiments with ligatures. Med. Rec. N. Y., 1867, ii., 449.

148. 1867. Porter. Dublin Med. Journal, 269.

149. 1867. N. Bubnoff. Ueber die Organization des Thrombus. Centralblatt. f. die Med. Wiss. No. 48. Arch. f. path. Anat. xliv., 1868.

150. 1867. C. Thiersch. Die feineren anatomischen Veränderungen nach Verwundung der Weichtheile. Handbuch der allgemeinen und speciellen Chirurgie von Pitha und Billroth, i., 531, Erlangen.

151. 1867. Simbert. Compt. rendus des Sci. et Mem. de la Société de Biologie, Paris.

152. 1867. G. W. Callender. Note on acupressure, Lancet, i., 597.

153. 1867. Conheim. Arch. f. Path. Anat. xl., 1 ; also Lectures on General Pathology.

154. 1867. Waldeyer. Zur pathologischen Anatomie der Wundkrankheiten, Arch. f. Path. Anat., xl., 301.

155. 1867. Henry Lee and Lionel S. Beale. Medico-Chirurg. Trans., i.

156. 1868. T. G. Morton. Review of the ligature of arteries at the Penn. Hosp. Penn. Hos. Rep., i., 192 ; also Am. J. M. Sci., 1876, n. s. lxxi., 334, lxxii. 17.

157. 1868. Hutchinson. The effect of acupressure on arteries, Med. Rec., N. Y. ii., 544.

158. 1868. A. Hewson. On acupressure. Penn. Hos. Rep. Phila. i., 127.

159. 1868. Thos. Bryant. Medico-Chir. Trans., xli.

160. 1868. Czerpay. Centralblatt f. die Med. Wissen., No. 1.

161. 1868. Max Bröer. Untersuchungen über die Organisation und Zerfall des Thrombus. Inaug. Diss., Breslau.

162. 1868. J. H. Pernet. De l'acupressure comme moyen hemostatique, Paris.

163. 1869. T. Waikloff. Das Gewebe des Ductus Arteriosus und die oblit. desselben, Ztschr. f. Rat. Med. Leipzig, xxxvi., 3 R. 109-131, 2 pl.

164. 1869. Pelechin. Studien über den Einfluss der entfernten Unterbindung von Hauptarterienstämmen auf die entsprechende Capillar. und Venen-circulation. Archiv. für. path. Anat., Berlin, xlv., 417.

165. 1869. Tschausoff. Ueber den Thrombus bei der Ligatur. Arch. f. klin Chir., xi., 184.

166. 1869. Theodor Kocher. Ueber die feineren Vorgänge bei der Blutstillungen durch Acupressur, Ligatur und Torsion. Arch. f. klin. Chir., xi., 660.

167. 1869. Richardson, Med. Times and Gazette, April 24.

168. 1869. G. A. Peters. An essay on acupressure with reference to its application in the continuity of arteries. N. Y. M. J., ix., 225.

169. 1869. Ogston. Lancet, April 17.

170. 1869. Pollock. New York Med. Jour.

171. 1869. G. M. Humphrey. On Torsion of arteries, with an account of some experiments. Brit. Med. Jour., London, i., 502.

172. 1870. F. Rizzoli. Sulla agopressione in ispecie pettrallamento d'alcune particolari cisti aneurismaticho. Mem. Acad. d'Sci. d. Inst. di Bologna, 2, s, x. 555.

173. 1870. Prengrueber. Sur l'acupressure dans le traitement des anévrysmes externes—Priorité des travaux du professeur Rizzoli (de Bologne) sur ceux de Simpson (d'Edinbourg) Gaz. hebd. de méd., Paris, 2s., vii., 420.

174. 1870. L. Ranvier. Epithelium. Nouveau Dict. de Méd. et de Chir. pratique.

175. 1870. Roser. Archiv. Zur Theorie der Blutstillung, Arch. f. Klin. Chir., xii., 1870.

176. 1870. S. Stricker. Ueber die Zelltheilung in entzündeten Geweben. Studien aus dem Institute f. experimentelle Path., Wein. Vorlesungen über allgemeinen und experimentelle Path., ii., 1878., Wein. Inter. Encyc. Surg., i., 1883.

177. 1871. R. Mayer. Lehrbuch der allg. path. Anat., Leipzig, 23.

178. 1871. Hanns Kundrat. Ueber die krankhaften Veränderungen der Endothelien. Wein, Med. Jahrb., ii.

179. 1871. Ercolani. Del processo anatomico di obliterazione delle arterie e della Vena ombilicale. Mem. d. Acad. d. sci., d' Inst. de Bologna, i.

180. 1871. F. Durante. Untersuchungen ueber die organization des Thromb. Wiener Med. Jahrb., 1872, 143. Untersuchungen ueber Entzündung der Gefässwände, ibid. 1871. 321. Arch. de Physiologie, 1872.

181. 1872. Albert Adamkiewicz. Die mechanischen Blutstillungsmittel bei verletzten Arterien von Paré bis auf die neueste zeit. Arch. f. klin. Chir., xiv.

182. 1872. Dudukaloff. Berträge zur Kenntniss des Verwachsungsprocesses unterbundener Gefässe, Wiener Med. Jahrb. ii.

183. 1872. C. A. Hart. Successful application of Dr. Speirs' artery constrictor. N. Y. Med. Jour. xv., 175-183 ; see also Tr. Med. Soc. N. Y. Albany, 1871, 269, 284.

184. 1873. Cornil et Ranvier. Manuel d'histologie pathologique, Paris.

185. 1873. E. Klein. Anatomy of the lymphatic system.

186. 1873. J. B. Ullersperger. Geschichtliche Berichtigungen über torsio arteriarum. Bl. f. Heilwissensch. München, iv., 17.

187. 1874. Nicolaus Stra.winsky. Ueber den Bau der Nabelgefässe und über ihren Verschluss nach der Geburt. Sitzungsberichte der kaiserl. Acadamie der Wissenschaften, lxx., Wein.

188. 1874. O. Heubner. Die luetische Erkrankung der Hirnarterien, Leipzig.

189. 1874. L. Szuman. Untersuchungen über den temporären und dauernden Verschluss der Gefäss lumina nach Unterbindung und Acupressur, Memoiren des Vereins russischer Aertze. Gekrönte Preisschrift der Breslauer Facultät. Centralblatt f. d. Med. Wiss., 1874, No. 49.

190. 1875. G. C. E. Weber. A new method of arresting hemorrhage. Med. Rec. N. Y. x., 305.

191. 1875. V. Czerny. Ein Aneurysma varicosum. Ein Beitrag zur Lehre von der Organisation geschichteter Thromben. Arch. f. path. Anat. lxii., 464.

192. 1875. B. Riedel. Die Entwickelung der Narbe im Blutgefäss nach der Unterbindung. Deutsche Zeitschrift für Chirurgie, vi., 459.

193. 1875. Paul Bruns. Die temporäre Ligatur der Arterien nebst einem Anhange über Lister's Catgutligatur. Zeitschrift f. Chir. v., 379.

194. 1875. F. Wilhelm Zahn. Untersuchungen über Thrombose. Bildung der Thromben, Arch. f. path. Anat. lxii., 81, ibid. lxix., 1884.

195. 1875. C. F. Maunder. Lancet. Lectures.

196. 1875. A. Verneuil. De la forcipressure. Bull. et Mém. Soc. de Chir. de Paris, n. s. 137.

197. 1875. Heinrich Haeser. Lehrbuch der Geschichte der Medicin und der epidemischen Krankheiten, Jena.

198. 1875. Köster. Ueber die Structur der Gefässwände und die Entzündung der Venen. Berliner klin. Woch. No. 43. Ueber Endarteritis und Arteritis. Berliner klin. Woch. 1876, No. 23.

199. 1876. Discussion du memoire de M. Tillaux sur la torsion des artères. Bull. et Mem. Soc. de Chir. de Par., n. s. ii., 277. Tillaux, ibid., 231.

200. 1876. Paul Baumgarten. Ueber die sogenannte Organisation des Thrombus, Centralb. f. d. med. Wiss., No. 34. Reprint 1877, Uber das offenbleiben fötaler Gefässe, Centlb. d. Med. Wis., No. 41., 1877.

201. 1876. Friedländer. Ueber Arteriitis obliterans. Centralb. f. d. medic. Wiss., No. 4.

202. 1876. R. McDonnell. Torsion of Arteries. Med. Press. and Circ., London, i., 153.

203. 1876. C. Faget. Obliteration of the Botal foramen and ductus arteriosus. N. Orl. M. & S. J., iv., 29–36.

204. 1876. Ernst Ziegler. Untersuchungen über pathologische Bindegewebs und Gefässneubildung, Würzburg.

205. 1877. Benjamin Auerbach. Ueber die Obliteration der Arterien nach Ligatur. Diss. Bonn.

206. 1877. Nadieschda Schultz. Ueber die Venarbung von Arterien nach Unterbindungen und Verwundungen. Diss. Leipzig.

207. 1878. F. V. Winiwarter. Endarteritis und Endophlebitis. Archiv. f. klin. Chir., 23.

208. 1878. Fritz Raab. Ueber die Entwickelung der Narbe im Blutgefäss nach der Unterbindung. Arch. f. klin. Chir., xxiii., 156; also Arch. f. Path. Anat., lxxv., 451–471. Berlin.

209. 1879. F. Ferriere. Historical researches on the ligature. St. Louis Clin. Rec. vi., 325.

210. 1879. M. Ramos. Estudio comparativo entre la forcipressura etc. Escula de Méd. Mexico, i., No. xiii., etc.

211. 1879. Pfitzer. Archiv. f. Path. Anat., Sept. 24.

212. 1879. Senftleben. Ueber den Verschluss der Blutgefässe nach der Unterbindung. Archiv. f. Path. Anat., Berlin, lxxvi., 421.

213. 1879. Edward O. Shakespeare. The nature of reparatory inflammation in arteries after ligature, acupressure, and torsion. The Toner Lectures, Washington, Allen's Human Anatomy, Philadelphia, 1883.

214. 1880. Arnaud. Contribution à l'étude de la ligature dans le traitement des anévrismes. Paris.

215. 1880. J. Böeckel. De la ligature antiseptique des gros troncs arteriels dans la continuité. Rev. Méd. de l'est. Nancy, xii., 203.

216. 1880. L. M. Reuss. De la ligature antiseptique des gros troncs arteriels. Jour. de thérap., Paris, vii., 542.

217. 1881. Servier. Déligation. Dict. encycl. de sci. Méd., Paris.

218. 1881. F. Treves. A case illustrating the condition of large arteries after ligature, etc. Proc. Roy. M. & Chir. Soc., London, ix., 25.

219. 1881. C. W. F. Uhde. Unterbindung von Arterien in der Continuität. Arch. f. klin. Chir. Berlin, xxvi., 840.

220. 1881. C. T. Dent. Med. Chir. Trans., lxiv., 64.

221. 1881. John Ashurst, Jr. Encyclopædia of Surgery, Vol. i., 554.

222. 1881. Lister. Clinical Societies Trans., xiv; also Lancet, 1881, i., 201, 275.

223. 1881–82. W. Greifenberger. Historisch-Kritische Darstellung der Lehre von der Unterbindung Blutgefässe, Zeitschrift f. Chirurg., Leipzig, xvi.

224. 1883. T. Holmes. British Med. Jour., June 9.

225. 1883. W. S. Walsham. British Med. Jour., Apr. 7.

226. 1883. Thoma. Arch. f. Path. Anat., xciii. and xcv., 1884.

227. 1883. J. Black. On the deligation of large arteries by the application of two ligatures and the division of the vessel between them. Brit. M. J., London, i., 765.

228. 1883. Richard Barwell. International Encyclopædia of Surgery, iii., New York. Also Med. Chir. Trans. lxiv., 1881.

229. 1883. John A. Liddell. International Encyclopædia of Surgery, iii., New York; also Med. Chir. Trans., lxiv., 1881.

230. 1883. Wyeth. Surgical Diseases of the Vascular System. Int. Encycl. Surg., iii., 351; also Med. Rec., N. Y., 1882, xxii., 103.

231. 1884. J. Collins Warren. The healing of arteries after ligature. Proceedings Boston Soc. Med. Sci., Mar. 20, 1883. Boston Med. and Surg. Jour., May 1, 1884, Philadelphia Med. Times, xvi. 132, 1885.

232. 1885. N. Senn. Cicatrization in blood-vessels after ligature. Trans. Am. Surg. Assoc., ii., Phila.

233. 1885. Fr. Burdach. Ueber den Senftlebenschen Versuch. die Bindegewebsbildung in todten, doppelt unterbundenen Gefässtrecken. betreffend. Arch. f. Path. Anat., c. 217, Berlin.

234. 1886. Experimentelle Untersuchungen über Thrombose. J. C. Eberth und C. Schinmelbusch, Archiv. f. Path. Anat., ciii., 39, Berlin.

235. 1886. William Osler. Cartwright Lectures. Med. News, xlviii., 690, 691, 692.

INDEX.

Artery.
 Paulus Ægineta, 4.
 femoral artery first tied at Poupart's ligament by Severinus, 8.
 contraction of arteries, important
  in arresting hemorrhage, 10.
 means of arresting hemorrhage in
  subcutaneous divisions, 11; in
  ordinary accidents, 11; in vessels partially divided, 12; in
  artery surrounded by tight
  ligature, 12.
 strength depends on its external
  coat, 12.
 agency of walls of the vessel in process of repair, 12.
 coagulum forms in wounds of
  arteries, 3.
 longitudinal slit may heal without
  obliteration of canal, 14.
 brachial artery of a stump, 22.
 narrowing of arteries in stump, 32.
 minute anatomy of the arteries, 45.
 time required for complete cicatrization, 97.
 appearances noticed in vessel of
  amputated stump, 100, 149.
 permanence of the umbilical artery, 129.
 Baumgarten maintains that an
  artery can be tied without formation of thrombus, 141.
 aorta, 112, 126, 127.
 axillary, 100.
 brachial, 22, 98.
 carotid, 50, 51, 52, 53, 70, 71, 76, 78,
  79, 83, 85, 91.
 cerebral, 26.
 ductus arteriosus, 32, 111, 113.
 femoral, 8, 48, 49, 53, 55, 57, 58, 59,
  62, 65, 67, 88, 99, 104.
 hypogastric (umbilical), 111, 113,
  129, 131, 132, 133.
 iliac, 89, 96.
 subclavian, 86, 95.
 tibial, 99, 100, 102, 103, 104, 106.
 umbilical, see hypogastric.
Asepsis.
 absence of thrombus with complete
  asepsis, 43, 44.
Avenzohar.
 duties of physicians, 5.
 use of ligatures, 5.
Averrhoes.
 use of ligatures, 5.
Avicenna.
 duties of physicians, 5.
 confines ligatures to arteries, 5.
Bandages applied about limb and provisional ligatures before amputation, 3.
Barwell.
 ox-aorta ligature, 39.

Baumgarten.
 observations on the growth of the
  thrombus, 23.
 growth of granulations, 23.
 behavior of the endothelial cells, 24.
 agreement with Pfitzer, 29.
 reply to Senftleben, 29.
 application of double ligature, 42.
 absence of thrombosis with strict
  asepsis, 43, 44.
 summary of views on formation of
  thrombus, 43.
 Virchow's axiom confirmed, 44.
 umbilical artery in the adult, 113.
 the ductus arteriosus an example
  of the "typical process of incomplete obliteration," 113.
 permanence of the umbilical artery,
  129.
 his view disproved, 135.
 maintains that an artery can be
  tied without formation of
  thrombus, 141.
 the extent of traumatism the most
  important factor in the production of the thrombus, 141.
Beale.
 arteries of horses and donkeys, 28.
 opening in elastic coat filled with
  colorless substance, 28.
 agency of white corpuscles, 28.
Bell.
 pressure exerted by surrounding
  tissue prevents hemorrhage,
  adhesive inflammation subsequently uniting the walls, 10.
Bertaplagia.
 transfixion of vessel, 6.
Billroth.
 adopts views of Bubnoff, 20.   [20.
 proliferation in white corpuscles,
Blandin.
 pronounces in favor of the organizing power of the thrombus, 13.
Bleeding—see hemorrhage.
Blood.
 action of blood in coagulating in
  arteries, 42.
 red corpuscles absorbed, 10.
 wall of the vessel, an important
  factor in preserving the fluidity
  of the blood, 44.
 action of blood between two ligatures, 12.
 plaques seen in dogs, 141.
Botalli.
 ductus Botalli, 15, 32.
Bouchon.
 Petit, 9,
 brachial, 9.
Bouillaud.
 pronounces in favor of organizing
  power of the thrombus, 12.

Index. 183

Torsion.
  Amussat the first to communicate a paper on torsion, 35.
  method employed by Velpeau, 35.
  comparison with the ligature and acupressure, 36.
  effect of torsion, 35.
  paper by Costello, 35.
  Galenus, 3.
  torsion as substitute for ligature, 152.
  Rufus of Ephesus and Heliodorus familiar with torsion, 3.
Tourniquet.
  invented by Morel, 8.
Transfixion of vessel, 6.
Tschausoff.
  investigations on brachial artery of a stump, 22.
  thrombus takes no part in the organization, 22.
  cicatrization accomplished by connective tissue elements, 22.
Ulceration of the ligature, 79.
Umbilical cord.
  ligature of umbilical cord, 1.
  action of the umbilical vessels, 10, 15, 32.
  umbilical (hypogastric) artery, 113.
  elastic lamina, 113.
  contraction of the media, 113.
  hyaline connective tissue, 113.
  Baumgarten's description of umbilical artery in the adult, 113.
  process of obliteration, 113.
  experiments on the fœtal vessels, 114, 129.
  peculiarities in anatomical structure, 124.
  earliest changes, 130.
  hyaline degeneration, 131, 135.
  conversion into cord, 131.
  growth into thrombus, 132.
  contraction of outer walls, 133.
  development of new tissue, 132.
  permanence of the umbilical artery, 129.
  muscular character of obliterating tissue, 134.
  thrombus found in new-born child, 134.
  lumen soon obliterated, 135.
  thrombus not of traumatic origin, 135.
  presence of large amount of longitudinal muscular fibre, 135.
  absence of well defined outline to inner wall of media, 135.
  contraction at time of birth, 135.
  permanence of vessel disproved, 135.
  growth of tissue at second month, 136.

Umbilical cord.
  contraction of coats, 136.
  growth of young cells, 136.
  granulation-like masses, 136.
  condition at adult age, 136.
  summary of changes since birth, 136.
  newly formed tissue of muscular character, 136.
  not developed from intima, 136.
  extreme end of the vessel not closed by fibrinous tissue, 137.
  experiments on the monkey, 137.
  calcification of the walls seen in tibial and femoral, not observed in hypogastric artery of adults, but found in ligamentum arteriosum, 137.
  resemblance to changes in arteries after amputation, 137.
  formation of new muscular fibre, 137.
  summary of results after closure, 151.
Unfolding of ends of vessel, 80.
  see opening.
Uterus.
  formation of muscular cells in walls of uterus, 148.
"Valet à Patin," 9.
Van Soligen.
  modification of the forceps, 8.
Vasa vasorum.
  action in causing growth of tissue, 10, 11. [46.
  presence in outer and middle coats,
Vascular spaces.
  regarded by Stilling as sinuses, 13.
Vascularization.
  of thrombus, 15, 16, 20, 24, 62.
Veins.
  difference between arteries and veins, 1.
Velpeau.
  experiments upon animals, 35.
  pressure sufficient to prevent bleeding, 35.
  circulation not under control of the heart, 35.
  method of performing torsion, 35.
Verneuil.
  narrowing of vessels, 33.
  contraction of the media, compensatory arteritis and atrophy of the muscular coat, 33.
  shape and size of the cicatrix, 33.
Vigo, de.
  accredited with the discovery of acupressure, 6.
Vinegar used to arrest hemorrhage, 2.
Virchow.
  omnis cellula e cellula, 16.
  action of the white corpuscles, 16.

184   *The Ligature of Arteries.*

Virchow.
stellate cells, 16.
sinus-like degeneration, 16.
vascularization of the thrombus,
16.
slowing of the blood-current as a
factor in coagulation, 42.
axiom concerning inflammation
and thrombosis, 44.
Waldeyer.
action of endothelium in producing
connective tissue, 19.
organization of the thrombus, 19.
Waller.
observations of passage of white
corpuscles, 20.
Walsham.
placed double ligature on the vessel, employing carbolized ligatures, 34.
Walther, von.
accepts Hunter's views, 13.
believes that the vessel-walls com-
bine with the clot to form a
cicatricial tissue, 13.
Wandering cells, 20, 21.   [31, 47.
action of wandering cells, 26, 29,
their passage through the walls,
49, 56, 64.
cells infiltrating walls, 109.
presence of wandering cells, 142.
in posterior tibial after three weeks,
104.

Warren Anatomical Museum.
specimens from, 83, 85, 91.
Weber.
confirms Virchow's views, 17.
organization of the thrombus, 17.
action of white corpuscles, 17.
stellate net-work, 18.
White.
opinion that clot prevents closure
of the artery, 10.
Winiwarter.
endarteritis, 24.
changes observed in the arteries
and veins, 26, 27.
formation of laminæ, 27.
minute anatomy of the arteries, 45.
Wiseman.
use of the " Royal Styptic," or the
cautery, 8.
Wyeth.
process observed in traumatic arteries, 28.
thinks division of the inner and
middle coats unnecessary, 33.
Zahn.
experiments on rabbits, 32.
development of the white thrombus, 43.
Zwicky.
repeats experiments of Jones and
Stilling, 14.
formation of vessels in the thrombus, 15.

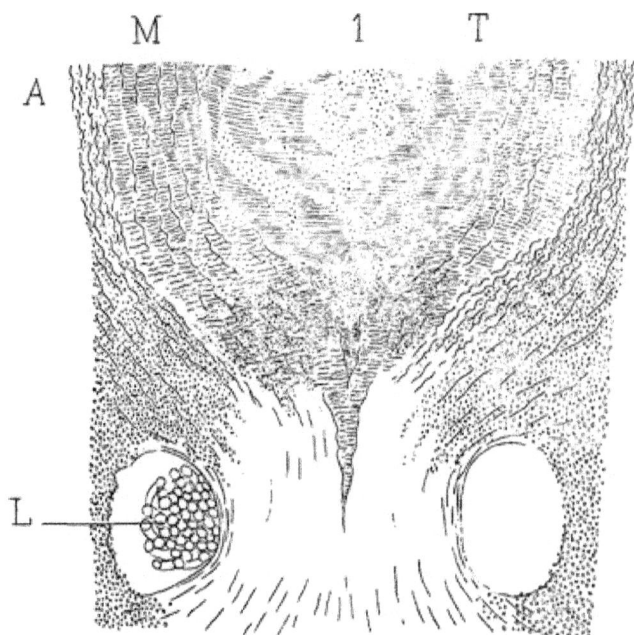

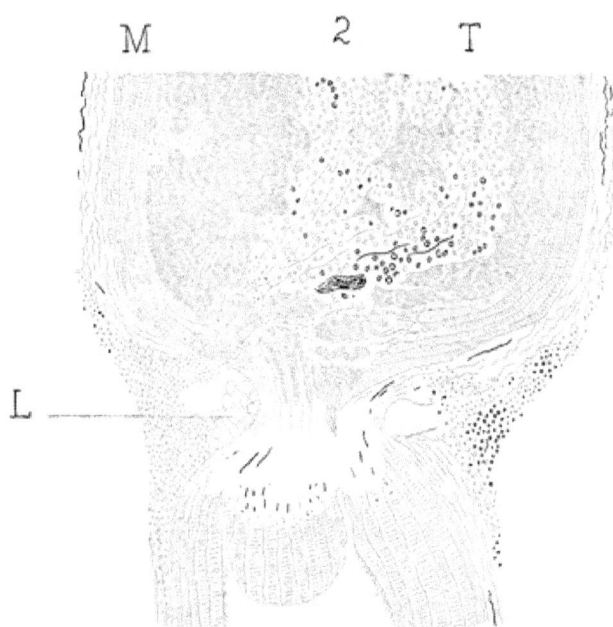

5

T

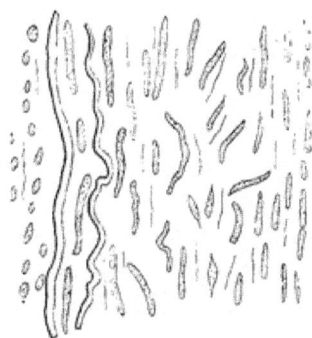

I

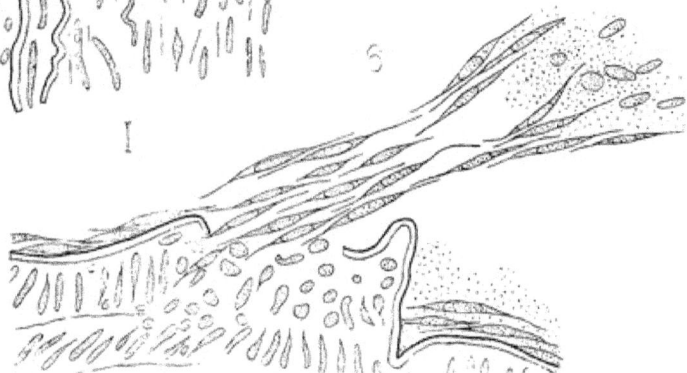

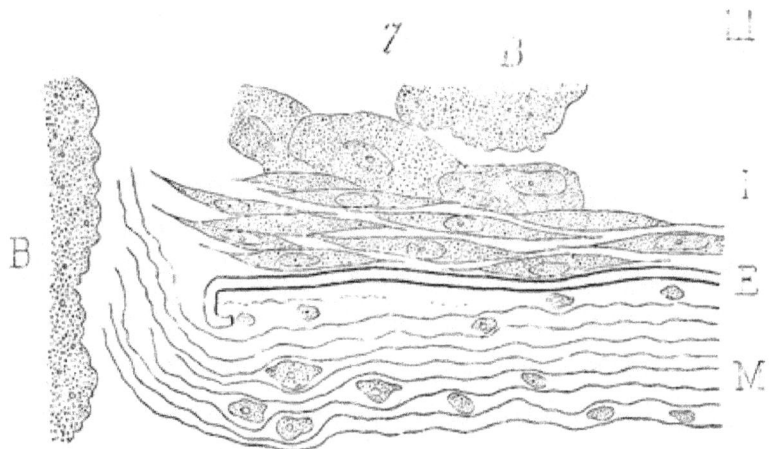

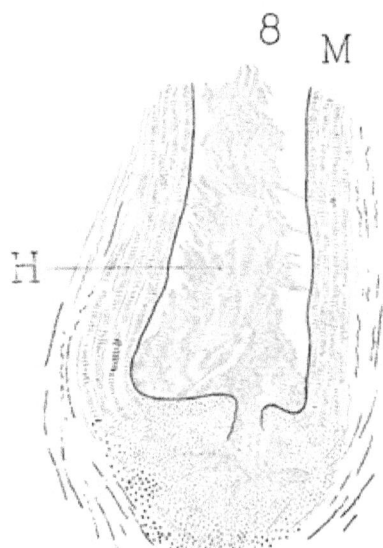

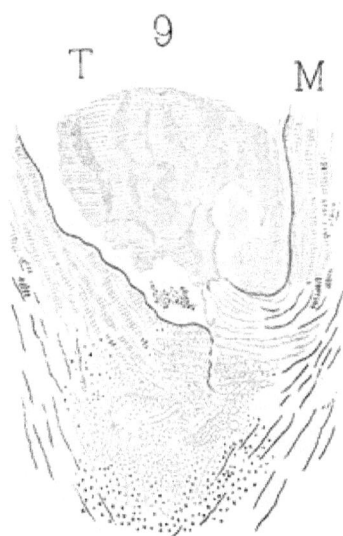

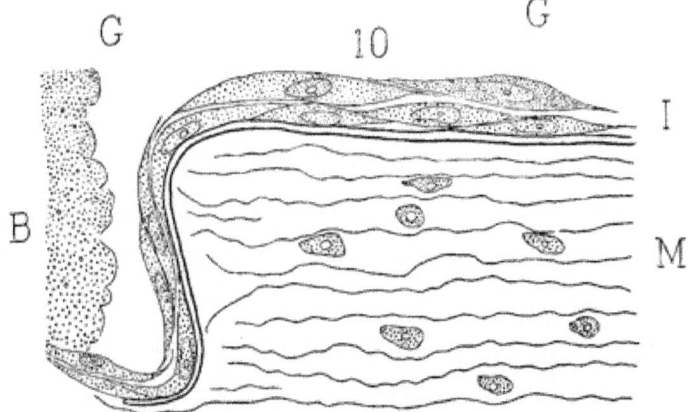

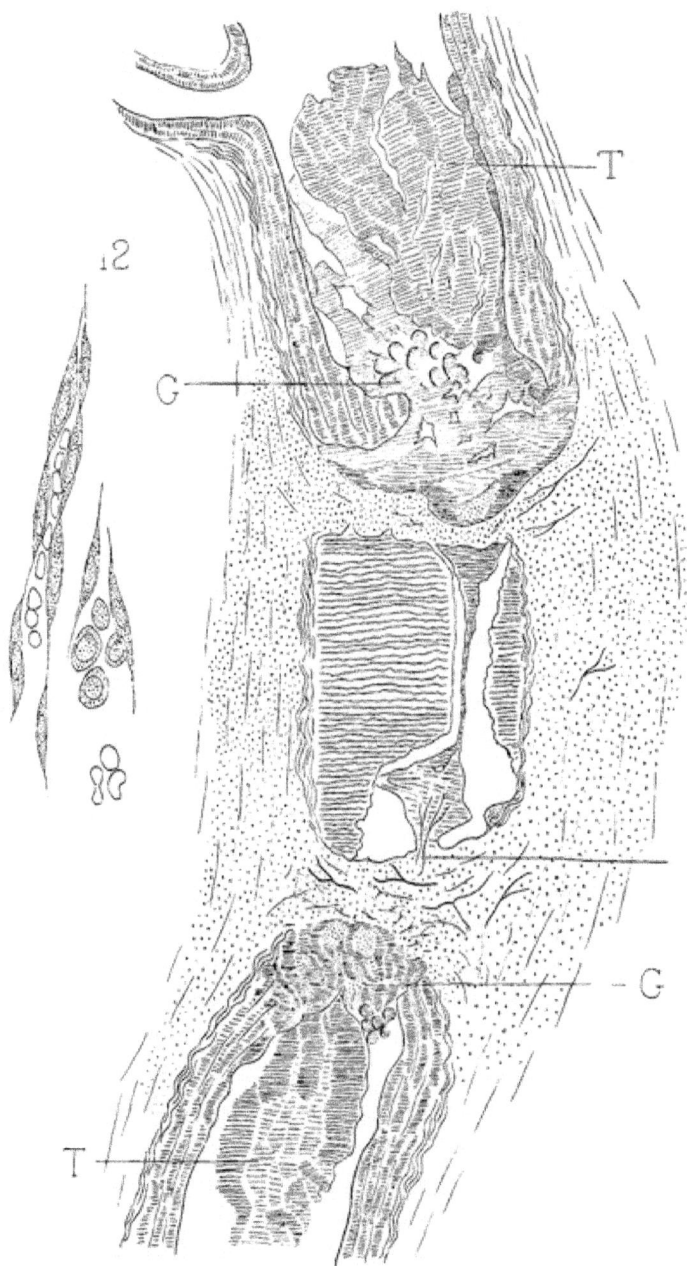

11

12

T

G

G

T

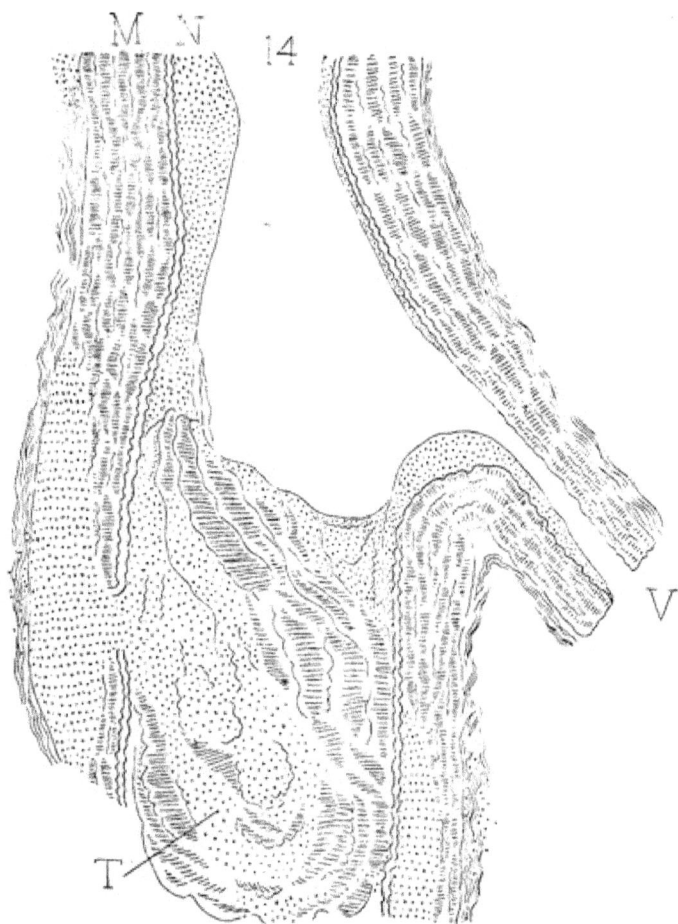

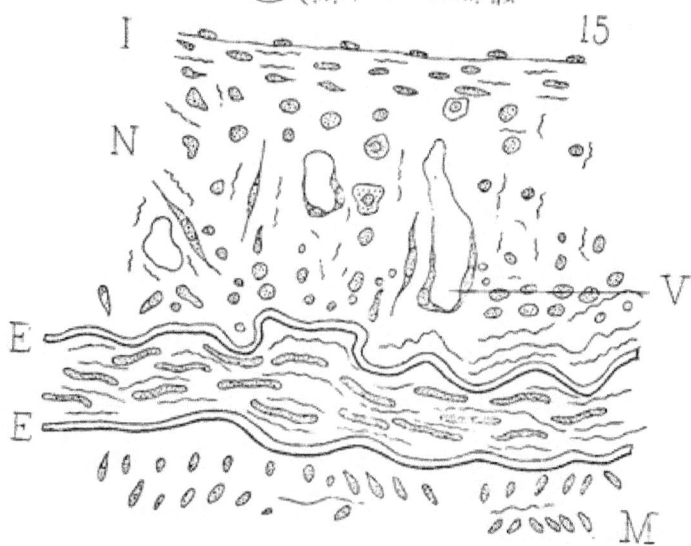

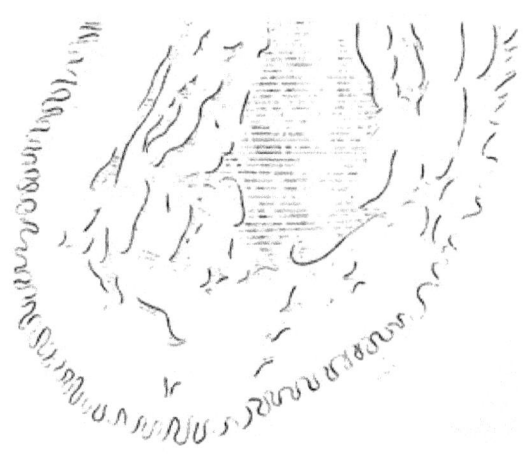

18

H

K

S

D

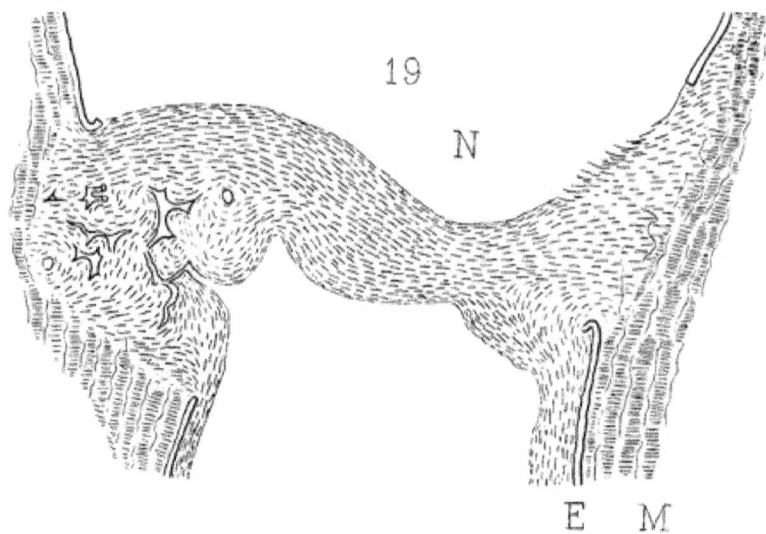

19

N

E M

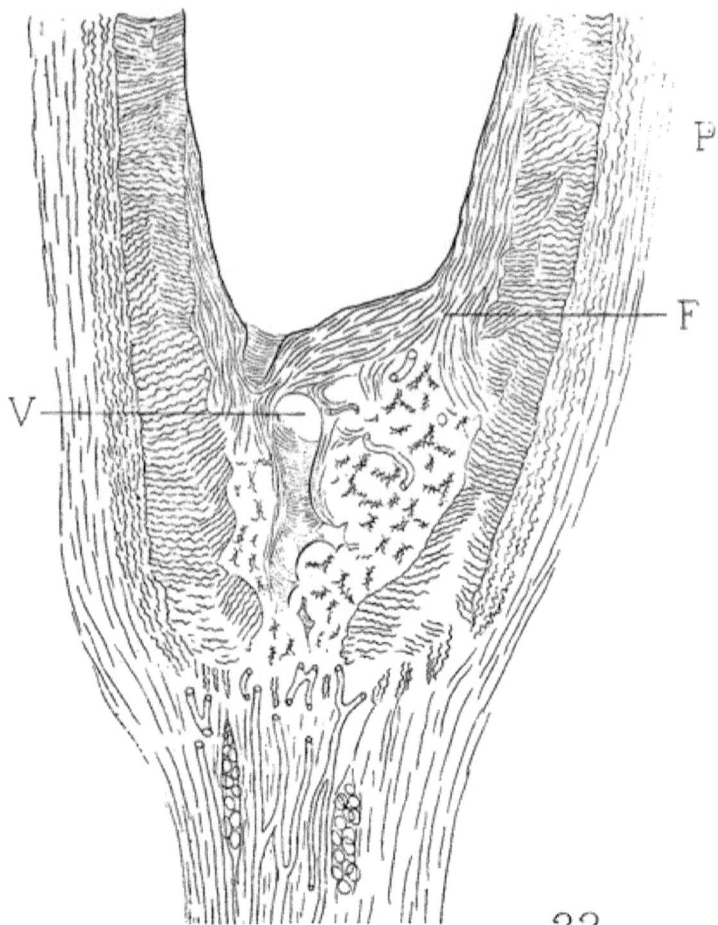

P

F

V

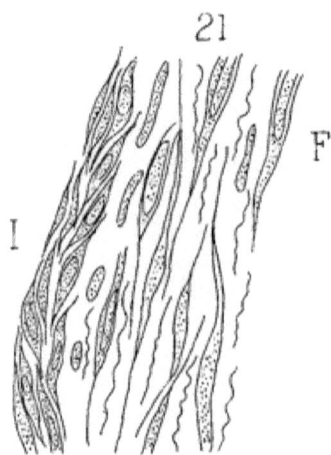

21

I

F

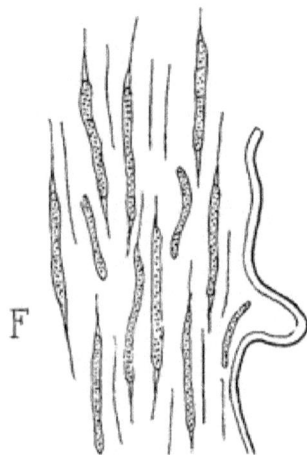

22

F

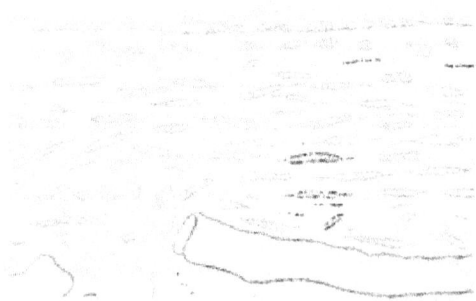

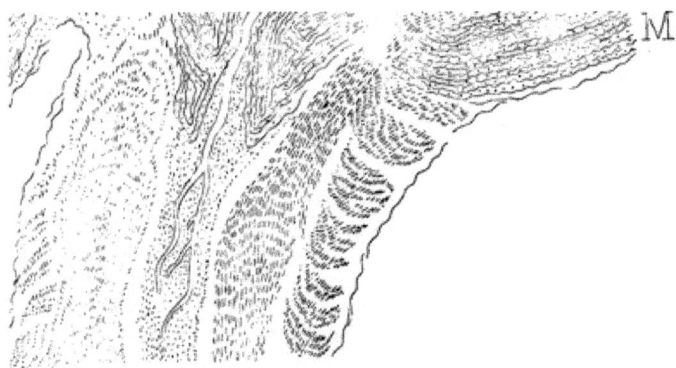

M

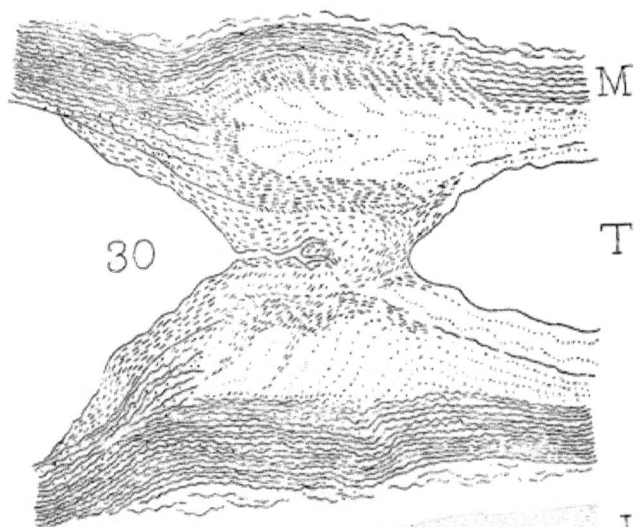

M

T

30

I

M

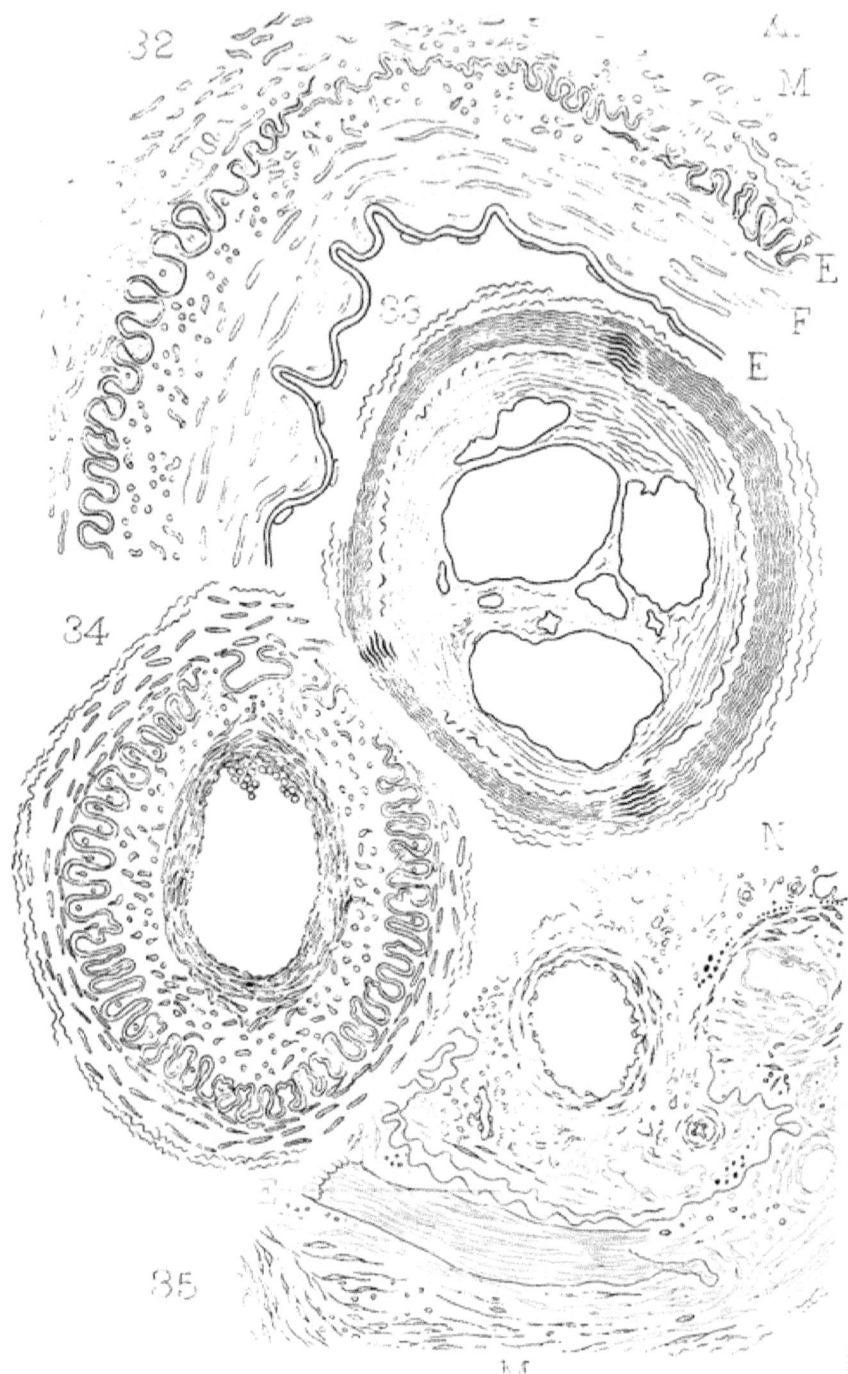